# Legal Issues of Renewable Energy in the Asia Region

Energy and Environmental Law & Policy Series Supranational and
Comparative Aspects

VOLUME 25

Editor

Kurt Deketelaere

*Professor of Law, University of Leuven, Belgium,
Honorary Chief of Staff, Flemish Government
Honorary Professor of Law, University of Dundee, UK
Secretary – General, League of European Research Universities (LERU), Belgium*

Editorial Board

*Dr Philip Andrews-Speed, Associate Fellow, Chatham House
Professor Michael Faure, University of Maastricht
Professor Günther Handl, Tulane University, New Orleans
Professor Andres Nollkaemper, University of Amsterdam
Professor Oran Young, University of California*

The aim of the Editor and the Editorial Board of this series is to publish works of excellent quality that focus on the study of energy and environmental law and policy.

Through this series the Editor and Editorial Board hope:

- to contribute to the improvement of the quality of energy/environmental law and policy in general and environmental quality and energy efficiency in particular;
- to increase the access to environmental and energy information for students, academics, non-governmental organizations, government institutions, and business;
- to facilitate cooperation between academic and non-academic communities in the field of energy and environmental law and policy throughout the world.

# Legal Issues of Renewable Energy in the Asia Region

Recent Developments in a Post-Fukushima and Post-Kyoto Protocol Era

Edited by

Anton Ming-Zhi Gao
Chien Te Fan

Law & Business

*Published by:*
Kluwer Law International
PO Box 316
2400 AH Alphen aan den Rijn
The Netherlands
Website: www.kluwerlaw.com

*Sold and distributed in North, Central and South America by:*
Aspen Publishers, Inc.
7201 McKinney Circle
Frederick, MD 21704
United States of America
Email: customer.service@aspenpublishers.com

*Sold and distributed in all other countries by:*
Turpin Distribution Services Ltd
Stratton Business Park
Pegasus Drive, Biggleswade
Bedfordshire SG18 8TQ
United Kingdom
Email: kluwerlaw@turpin-distribution.com

*Printed on acid-free paper.*

ISBN 978-90-411-4856-8

© 2014 Kluwer Law International BV, The Netherlands

All rights reserved. No part of this publication may be reproduced, stored in a retrieval system, or transmitted in any form or by any means, electronic, mechanical, photocopying, recording, or otherwise, without written permission from the publisher.

Permission to use this content must be obtained from the copyright owner. Please apply to: Permissions Department, Wolters Kluwer Legal, 76 Ninth Avenue, 7th Floor, New York, NY 10011-5201, USA. Email: permissions@kluwerlaw.com

Printed and Bound by CPI Group (UK) Ltd, Croydon, CR0 4YY.

# List of Editors and Contributors

**Dr. Chien Te Fan**, Professor and Director in the Institute of Law for Science and Technology at the National Tsing Hua University (Taiwan). He specializes in energy and environmental law, areas of biotechnology, ethic related issues and competition law. He is also an experienced scholar involving in UNFCCC negotiation process since 2001 in Marrakesh. Moreover, as a speaker, he was also invited to join the side event several timesheld in the COPs.

**Dr. Anton Ming-Zhi Gao**, Assistant Professor in the Institute of Law for Science and Technology at the National Tsing Hua University (Taiwan). Main activities are concentrated in the areas of energy law and policy, European environmental law, renewable energy, feed in tariff, strategic environmental assessment, etc.

**Dr. Madjedi Hasan**, Director of PT Pranata Energi Nusantara Consulting, an energy development and specific engineering process consultant. He is also a fellow charter arbitrator in Indonesia National Arbitration Institute. He holds Master of Petroleum Engineering degree, and Master and Doctor in Business Law, and has more than 50 years of experiences in Indonesia's petroleum and geothermal power projects.

**Dr. Popi Konidari** is the Head of the Climate Change Policy Unit of the Energy Policy and Development Centre (KEPA) of the National and Kapodistrian University of Athens. She has an MSc in environmental physics and an MSc in electronic automatisms. She is currently working as a senior researcher at KEPA on climate change policy issues (emission trading schemes, evaluation criteria for climate change policy instruments, policy interactions, AMS method, development of mitigation/adaptation policy portfolios, Kyoto mechanisms). She is the scientific secretariat of the bi-lingual scientific journal 'Euro- Asian Journal of Sustainable Energy Development Policy'.

**Dr. Taehwa Lee**, a Research Professor of the Institute for Legal Studies at Yonsei University. She received her Ph.D. in Urban Affairs and Public Policy from University of Delaware, USA. Her research interests are environmental & energy policy and political economy of Climate Change regime in relation to global trade and investment regimes.

List of Editors and Contributors

**Dr. Ken'ichi Matsumoto** is an Assistant Professor at the School of Environmental Science, The University of Shiga Prefecture. He received his Ph.D. in Policy Studies from Kwansei Gakuin University in 2007. He specializes in environmental/energy policy and economics.

**Dr. Kanako Morita** is a Project Assistant Professor at Graduate School of Media and Governance, Keio University. She received her Ph.D. in Value and Decision Science from Tokyo Institute of Technology in 2010. She specializes in environmental policy and governance, including environmental financing.

**Dr. Deok-Young Park** is a Professor of International Economic Law at Yonsei University. He finished Ph.D coursework in Edinburgh University and received his Ph.D. in law from Yonsei University. He specializes in International Dispute Settlement, Trade and Environment and Trade Related Intellectual Properties.

**Jingli Shi**, Professor of Energy Research Institute and CNREC, graduated from Tsinghua University and North China Institute of Electric Power with bachelor and master degrees, and since 1995, she has engaged in the economic assessment and policy research in the field of renewable energy, especially electricity pricing policy and involved in drafting a number of important national policy documents.

**Darryl Smith**, Independent consultant in the electricity and telecommunications industries in Australia. He holds a degree in Electrical Engineering and has spent 12 years with a state-owned generation company in both field operations and head office functions. He has also assisted a private company to enter the Smart Grid market.

**Nucharee Nuchkoom Smith** is a Doctoral student, School of Law, at the University of Western Sydney, Australia specializing in International Trade Law. She holds Masters Degrees from Universities in both Thailand and Australia. She was previously a Legal Officer to the House of Representatives Standing Committee of Foreign Affairs of the Thai Parliament.

**Dr. Robert Smith** holds a Ph.D. in Civil Engineering and is currently a Team Leader for an Asian Development Bank funded asset management advisory project for Karnataka Public Works, Ports and Inland Water Transport Department India in India. He specializes in the field of infrastructure asset management and financial sustainability of asset management practices.

**Manuel Solis** is a Teaching Fellow at the Law School, University of Adelaide. He specializes in energy, climate change, environment and sustainable development law. He has extensive work experience on institutional capacity, regulatory and policy framework development projects involving government, business, non-government organizations, and multilateral and bilateral development agencies such as the UNDP, World Bank and the Australian Agency for International Development, among others.

**Antonius Wahjosoedibjo**, President Director of PT Pranata Energi Nusantara Consulting. He is a graduate electrical engineer and has spent almost 40 years of his career with Chevron's affiliated companies in Indonesia and the United States. He is a member of the Board of Advisors and/or the Board of Experts of the Indonesian Electrical Power

Society (MKI), the Indonesian Renewable Energy Society (METI), the Indonesian Geothermal Association (API), and the Indonesian Power Engineers Association (IATKI).

**Dr. Tao Ye**, Researcher of ERI and CNREC, graduated from Department of Automation of Northwestern Polytechnical University with Ph.D. in 2009. He has engaged in renewable energy modeling research and involved in research of RE Power Quota System of China.

**Jingting Yuan**, China National Renewable Energy Center researcher of CNREC, graduated from Tsinghua University of Management Science and Engineering with master. She has engaged in wind power policy research and RE policy research.

# Summary of Contents

| | |
|---|---:|
| List of Editors and Contributors | v |
| Preface | xxi |
| Acknowledgement | xxv |

PART I
New Renewable Electricity Promotion Regime after the Fukushima Accident     1

CHAPTER 1
Renewable Energy-Related Policies and Institutions in Japan: Before and after the Fukushima Nuclear Accident and the Feed-In Tariff Introduction
*Kanako Morita & Ken'ichi Matsumoto*     3

CHAPTER 2
From FIT to RPS under the Low-Carbon Green Growth Initiative: Moving Forward or Backward for the Expansion of Renewable Energy in Korea?
*Deok-Young Park & Taehwa Lee*     29

CHAPTER 3
A More Sustainable Way to Promote PV: Transformations from FIT to FIT/FIT Tendering Schemes in Taiwan and France
*Anton Ming-Zhi Gao*     47

PART II
The Evolution of the Existing Renewable Electricity Promotion Scheme after Fukushima Accident     85

Summary of Contents

CHAPTER 4
Crossroad of FIT and RPS: What's the Next Step for China?
*Jingli Shi, Tao Ye & Jingting Yuan*     87

CHAPTER 5
Renewable Energy Development in the Philippines: Legal Measures, Implementation, Challenges, and Solutions
*Manuel Peter S. Solis*     103

CHAPTER 6
FIT and Its Implementation in Thailand: Legal Measures, Implementation, Challenges, and Solutions
*Robert Brian Smith, Nucharee Nuchkoom Smith & Darryl Robert Smith*     127

CHAPTER 7
Feed-In Tariff for Indonesia's Renewable Electricity
*Madjedi Hasan & Anton S. Wahjosoedibjo*     147

PART III
Cross Country Analysis of Renewable Electricity Promotion Regime     161

CHAPTER 8
Evaluation of Eleven Implemented Policy Mixtures in the Black Sea and Caspian Sea Regions for the Use of RES
*Popi Konidari*     163

CHAPTER 9
Transformation of German- and European-Style Feed-In Tariff Schemes in East Asia in the Post-Fukushima Age: Recent Developments in Japan, South Korea, and Taiwan
*Anton Ming-Zhi Gao & Chien Te Fan*     213

# Table of Contents

| | |
|---|---|
| List of Editors and Contributors | v |
| Preface | xxi |
| Acknowledgement | xxv |

PART I
New Renewable Electricity Promotion Regime after the Fukushima Accident — 1

CHAPTER 1
Renewable Energy-Related Policies and Institutions in Japan: Before and after the Fukushima Nuclear Accident and the Feed-In Tariff Introduction
*Kanako Morita & Ken'ichi Matsumoto* — 3

| | | | |
|---|---|---|---|
| §1.01 | Introduction | | 3 |
| §1.02 | A Brief Overview of Renewable Energy-Related Policies and Institutions in Japan before the Introduction of the FIT Scheme in 2012 | | 5 |
| | [A] | Renewable Energy-Related Policies and Legal Measures | 5 |
| | [B] | Implementation and Effects of Major Renewable Energy-Related Schemes | 7 |
| | [C] | Challenges Facing Major Renewable Energy-Related Schemes | 9 |
| §1.03 | FIT Scheme in Japan | | 10 |
| | [A] | Reasons for Introducing the FIT Scheme | 11 |
| | [B] | Architecture of the FIT Scheme | 11 |
| | | [1] Overview of the Process of the FIT Scheme | 11 |
| | | [2] Purchase Prices and Periods under the FIT Scheme | 12 |
| | | [3] Certification of Producers of Renewable Electricity | 14 |
| | | [4] Purchase of Electricity by Electricity Companies | 15 |
| | | [5] Surcharge Adjustment under the FIT Scheme | 15 |

Table of Contents

|  |  | [6] | Revision of the FIT Scheme | 16 |
|---|---|---|---|---|
|  | [C] |  | Implementation and Effects of the FIT Scheme in Japan | 16 |
|  |  | [1] | Production of Renewable Electricity under the FIT Scheme | 16 |
|  |  | [2] | Current Situation of Renewable Energy Facilities | 18 |
|  | [D] |  | Challenges Facing the FIT Scheme and Improvements to the Scheme | 19 |
|  |  | [1] | Purchase Price and Long-Term Goal | 20 |
|  |  | [2] | Grid Connection | 20 |
|  |  | [3] | Procedures | 22 |
|  |  | [4] | Institutions and Ministries regarding Renewable Energy | 22 |
| §1.04 |  |  | Concluding Remarks | 28 |

CHAPTER 2
From FIT to RPS under the Low-Carbon Green Growth Initiative: Moving Forward or Backward for the Expansion of Renewable Energy in Korea?
*Deok-Young Park & Taehwa Lee*     29

| §2.01 |  | Introduction |  | 29 |
|---|---|---|---|---|
| §2.02 |  | Literature Review on FIT and RPS Programs |  | 31 |
| §2.03 |  | The FIT Program in Korea |  | 32 |
| §2.04 |  | The RPS Program in Korea since 2012 |  | 35 |
|  | [A] | The General Structure of the RPS Program |  | 35 |
|  | [B] | Issuance and Trading of REC |  | 37 |
| §2.05 |  | Challenges and Solutions |  | 39 |
|  | [A] | Challenges |  | 39 |
|  | [B] | Solutions: How to Improve the Effectiveness of RPS |  | 40 |
|  |  | [1] | Multiplier and Public Participation | 40 |
|  |  | [2] | Eligible Renewables | 43 |
|  |  | [3] | Dual Track System of FIT and RPS | 44 |
|  |  | [4] | Linking the RPS and ETS | 44 |
| §2.06 |  | Conclusion |  | 45 |

CHAPTER 3
A More Sustainable Way to Promote PV: Transformations from FIT to FIT/FIT Tendering Schemes in Taiwan and France
*Anton Ming-Zhi Gao*     47

| §3.01 |  | Introduction |  | 48 |
|---|---|---|---|---|
| §3.02 |  | Tendering Schemes in France |  | 50 |
|  | [A] | Background: The Adoption of FIT and Tendering Schemes in 2000 |  | 50 |
|  | [B] | Combo Tendering Scheme and FIT in 2011 and 2013 |  | 51 |

|     | [C] | Simple Tendering Scheme in 2011 and 2013 | 52 |
|---|---|---|---|
|     |     | [1] Policy Targets | 52 |
|     |     | [2] Tendering Targets and Caps | 53 |
|     |     | [3] Eligibility | 53 |
|     |     |     [a] Installation Capacity | 53 |
|     |     |     [b] Capacity Requirements for Each Case | 54 |
|     |     |     [c] Other Criteria | 54 |
|     |     | [4] The Ratio of Rooftops and Ground PV | 55 |
|     |     | [5] Administrative Procedures for Tendering | 55 |
|     |     |     [a] Submission of Related Documents | 55 |
|     |     |     [b] Review Process | 56 |
|     |     | [6] Criteria Required to Win Bids | 56 |
|     |     | [7] Procedures to Be Followed after Bids Are Won; Avoiding Delays in Construction Clauses | 57 |
|     |     | [8] Results of Tendering: Five Results | 58 |
|     | [D] | 2011 and 2013 Complex Tendering Schemes | 58 |
|     |     | [1] Policy Targets | 58 |
|     |     | [2] Tendering Targets and Caps | 59 |
|     |     | [3] Eligibility | 60 |
|     |     |     [a] Installation Capacity | 61 |
|     |     |     [b] Capacity Requirements for Each Case | 61 |
|     |     |     [c] Other Criteria | 61 |
|     |     | [4] Ratios of Rooftops and Ground PV | 65 |
|     |     | [5] Administrative Procedures for Tendering | 65 |
|     |     |     [a] Submission of Related Documents | 65 |
|     |     |     [b] Review Process | 66 |
|     |     | [6] Criteria to Win Bids | 66 |
|     |     | [7] Procedures to Be Followed after Bids Are Won; Avoiding Delays in Construction Clauses | 68 |
|     |     | [8] Results of Tendering | 69 |
| §3.03 | Tendering Scheme in Taiwan | | 71 |
|     | [A] | Background: The Adoption of FIT in 2009 | 71 |
|     | [B] | The Adoption of a Combo Scheme of FIT and Tendering in 2011 | 71 |
|     | [C] | Tendering Scheme | 72 |
|     |     | [1] Policy Targets | 72 |
|     |     |     [a] Prior to the 2011 Fukushima Accident | 72 |
|     |     |     [b] After the 2011 Fukushima Accident | 73 |
|     |     | [2] Tendering Targets and Caps | 74 |
|     |     | [3] Eligibility | 76 |
|     |     |     [a] Installation Capacity | 76 |
|     |     |     [b] Capacity Requirements for Each Case | 76 |
|     |     |     [c] Other Criteria | 76 |

Table of Contents

|  |  | [4] | Ratios of Rooftop and Ground-Type PV | 77 |
|---|---|---|---|---|
|  |  | [5] | Administrative Procedures for Tendering | 77 |
|  |  |  | [a] Submission of Related Documents | 77 |
|  |  |  | [b] Review Process | 78 |
|  |  | [6] | Criteria Required to Win Bids | 79 |
|  |  | [7] | Procedures to Be Followed after Bids Are Won; Avoiding Delays in Construction Clauses | 80 |
|  |  | [8] | Tendering Results: 18 Events | 80 |
| §3.04 | Conclusion |  |  | 81 |

PART II
The Evolution of the Existing Renewable Electricity Promotion Scheme after Fukushima Accident ........ 85

CHAPTER 4
Crossroad of FIT and RPS: What's the Next Step for China?
*Jingli Shi, Tao Ye & Jingting Yuan* ........ 87

| §4.01 | Introduction |  |  | 87 |
|---|---|---|---|---|
| §4.02 | Feed-In Tariff in China |  |  | 88 |
|  | [A] | The Detailed Design of the Feed-In Tariff Scheme |  | 88 |
|  |  | [1] | Technology Eligibility | 88 |
|  |  | [2] | FIT Duration | 90 |
|  |  | [3] | Tariff | 90 |
|  |  |  | [a] Tariff Schedule | 90 |
|  |  | [4] | Capacity Cap | 92 |
|  |  | [5] | Loading Hours (Resources Quality Cap) | 92 |
|  |  | [6] | Cost Sharing and Recovery | 92 |
|  |  | [7] | Grid Connection, Usage, and Expansion Rules | 93 |
|  | [B] | The Results of the FIT: Implementation |  | 94 |
|  | [C] | Challenges and Solutions |  | 94 |
| §4.03 | The Next Step: The Recent Development of RPS in China |  |  | 95 |
|  | [A] | Background of China's RPS Design |  | 95 |
|  | [B] | The Detailed Design of the RPS |  | 96 |
|  |  | [1] | Main Reference | 97 |
|  |  | [2] | Basic Principles | 97 |
|  |  | [3] | Development Target | 98 |
|  |  | [4] | Technology Scope | 99 |
|  |  | [5] | Duty-Bearers | 99 |
|  |  | [6] | Regulatory Agencies | 100 |

|  |  |  |
|---|---|---|
| | [C] Challenges and Solutions | 100 |
| §4.04 | Conclusion | 100 |

CHAPTER 5
Renewable Energy Development in the Philippines: Legal Measures,
Implementation, Challenges, and Solutions
*Manuel Peter S. Solis*                                                                 103

|  |  |  |
|---|---|---|
| §5.01 | Introduction | 103 |
| | [A] The Energy Situation | 104 |
| | [B] Renewable Energy Sources and the National Renewable Energy Program | 105 |
| | [1] Solar | 106 |
| | [2] Biomass | 107 |
| | [3] Hydropower | 107 |
| | [4] Wind | 108 |
| | [5] Ocean | 108 |
| §5.02 | The Renewable Energy Act of 2008 | 109 |
| | [A] RPS | 110 |
| | [B] FIT | 110 |
| | [C] Green Energy Option | 111 |
| | [D] Net-Metering | 111 |
| | [E] Fiscal Incentives | 112 |
| §5.03 | The Feed-In-Tariff Scheme | 112 |
| | [A] Coverage | 112 |
| | [B] Duration | 113 |
| | [C] Installation Target, FIT Rate, and Degression Rate | 113 |
| | [D] Cost-Sharing, Settlement, and FIT Adjustments | 114 |
| | [E] Priority Connection, Purchase, and Transmission | 115 |
| | [F] Administration and Review | 115 |
| §5.04 | The Proposed RPS Scheme | 116 |
| | [A] Eligible Renewable Energy Technologies | 116 |
| | [B] Mandated Participant | 117 |
| | [C] Compliance Mechanism | 118 |
| | [D] Establishment of the Renewable Energy Market and the Renewable Energy Registrar | 119 |
| | [E] Minimum Annual RPS Requirement and Annual Increment | 119 |
| §5.05 | The Implementation Challenges to the FIT Scheme | 120 |
| | [A] Concern on a Customer-Based FIT | 120 |
| | [B] Issue on Fit Entitlement | 122 |
| | [C] FIT Uncertainty upon Full Subscription of Installation Target | 123 |
| §5.06 | Conclusion | 124 |

## Table of Contents

CHAPTER 6
FIT and Its Implementation in Thailand: Legal Measures, Implementation, Challenges, and Solutions
*Robert Brian Smith, Nucharee Nuchkoom Smith & Darryl Robert Smith*    127

| | | |
|---|---|---|
| §6.01 | Introduction | 127 |
| | [A] Overview | 127 |
| | [B] Energy Industry Act BE 2550 (2007) | 129 |
| §6.02 | Feed-In Tariff Scheme | 131 |
| | [A] The Detailed Design of the Feed-In Tariff Scheme | 131 |
| |    [1] Technology Eligibility | 131 |
| |    [2] FIT Duration | 133 |
| |    [3] Tariff | 133 |
| |       [a] Tariff Schedule | 133 |
| |       [b] Tariff Degression Mechanism | 135 |
| |       [c] Tariff Progression Mechanism | 135 |
| |    [4] Capacity Cap | 135 |
| |       [a] Soft Cap | 135 |
| |       [b] Hard Cap | 136 |
| |    [5] Loading Hours (Resources Quality Cap) | 136 |
| |       [a] Small Power Producers | 136 |
| |       [b] Very Small Power Producers | 137 |
| |    [6] Cost Sharing and Recovery | 137 |
| |    [7] Grid Connection, Usage, and Expansion Rules | 138 |
| | [B] The Results of the FIT Implementation | 139 |
| §6.03 | Recent Discussion over FIT at a Post-Fukushima and Post-Kyoto Protocol Era | 142 |
| | [A] Evaluation | 142 |
| | [B] Challenges | 143 |
| | [C] Solutions | 145 |
| §6.04 | Conclusion | 146 |

CHAPTER 7
Feed-In Tariff for Indonesia's Renewable Electricity
*Madjedi Hasan & Anton S. Wahjosoedibjo*    147

| | | |
|---|---|---|
| §7.01 | Introduction | 147 |
| §7.02 | Energy Diversification | 148 |
| §7.03 | FIT Scheme | 150 |
| §7.04 | Overview of FIT Policies in Indonesia | 152 |
| | [A] Geothermal | 153 |
| | [B] Other Renewable Energies | 155 |
| | [C] Duration of FIT | 156 |
| | [D] Price Escalation | 157 |

|  |  | [E] Key Input Parameters | 158 |
| --- | --- | --- | --- |
| §7.05 |  | Implementation of FIT | 158 |
| §7.06 |  | Other Barriers to RE Development | 159 |
| §7.07 |  | Conclusion | 160 |

PART III
Cross Country Analysis of Renewable Electricity Promotion Regime      161

CHAPTER 8
Evaluation of Eleven Implemented Policy Mixtures in the Black Sea and
Caspian Sea Regions for the Use of RES
*Popi Konidari*      163

| §8.01 | Introduction |  | 163 |
| --- | --- | --- | --- |
| §8.02 | Policy Mixtures for the Use of RES |  | 165 |
|  | [A] | Objectives | 165 |
|  | [B] | Policy Instruments and Implementation Network | 166 |
|  | [C] | Albania | 167 |
|  | [D] | Armenia | 167 |
|  | [E] | Azerbaijan | 168 |
|  | [F] | Bulgaria | 168 |
|  | [G] | Estonia | 169 |
|  | [H] | Moldova | 170 |
|  | [I] | Romania | 171 |
|  | [J] | Russia | 172 |
|  | [K] | Serbia | 174 |
|  | [L] | Turkey | 174 |
|  | [M] | Ukraine | 175 |
| §8.03 | LEAP Model |  | 176 |
|  | [A] | Use of LEAP | 176 |
| §8.04 | AMS Method |  | 177 |
| §8.05 | Evaluation |  | 179 |
|  | [A] | Results | 204 |
| §8.06 | Conclusions |  | 204 |

References      204

CHAPTER 9
Transformation of German- and European-Style Feed-In Tariff Schemes
in East Asia in the Post-Fukushima Age: Recent Developments in Japan,
South Korea, and Taiwan
*Anton Ming-Zhi Gao & Chien Te Fan*      213

§9.01    Introduction      213

## Table of Contents

| | | | | |
|---|---|---|---|---|
| §9.02 | | | elopment of Primary Renewable Electricity Promotion Schemes in<br>an, South Korea, and Taiwan | 215 |
| | [A] | Japan | | 215 |
| | | [1] | Phase I: Preparatory Stage: Prior to 1997 | 215 |
| | | [2] | Phase II: 1997–2002 | 215 |
| | | [3] | Phase III: RPS Law and Voluntary<br>Net-Metering Scheme | 217 |
| | | | [a] RPS Law in 2003 | 217 |
| | | | [b] Voluntary Net-Metering Scheme | 220 |
| | | | [c] Other Incentives | 220 |
| | | [4] | Phase IV: 2009 Mandatory PV Net-Metering | 222 |
| | | | [a] PV Net-Metering Scheme | 222 |
| | | | [b] Other Incentives | 223 |
| | | | [c] Passing the Costs on to Consumers | 224 |
| | | [5] | Phase V: Post-Fukushima: 2011 FIT with Mandatory<br>Small PV Net-Metering | 225 |
| | | | [a] Detailed Comparison with PV Net-Metering<br>Scheme | 226 |
| | | | [b] Results | 229 |
| | | | [c] Potential Challenges | 229 |
| | [B] | South Korea | | 229 |
| | | [1] | Phase I. Soft Law in the 1990s | 230 |
| | | [2] | Phase II: 2000s | 230 |
| | | | [a] Feed-In Tariffs | 230 |
| | | | [b] Other Supplementary Schemes | 236 |
| | | [3] | Phase III. 2012: RPS and Mandatory Capacity<br>Installation of PV | 238 |
| | | | [a] RPS | 238 |
| | | | [b] Mandatory Capacity Installation of PV | 241 |
| | | | [c] Other Measures | 241 |
| | [C] | Taiwan | | 242 |
| | | [1] | Phase I: Before Drafting Renewable Energy Bill in 2002 | 242 |
| | | [2] | Phase II: 2003 Restricted Version of FIT: Transition<br>Period between 2002 and 2009 | 243 |
| | | [3] | Phase III: 2009 Renewable Energy Act: FIT – Adoption of<br>Renewable Energy Act in 2009 | 247 |
| | | [4] | Phase IV: Post-PV Boom and Post-Fukushima: New<br>Energy Policy of November 2011 – FIT + PV Tendering | 249 |
| | | | [a] Eligibility | 249 |
| | | | [b] Duration | 250 |
| | | | [c] Rate Schedule | 250 |
| | | | [d] Development Target and Cap | 252 |

| | | | | |
|---|---|---|---|---|
| §9.03 | Analysis: Unique Schemes in Japan, South Korea, and Taiwan | | | 254 |
| | [A] | Comparison with German-Style FIT: Unique FITs in Japan, South Korea, and Taiwan | | 254 |
| | | [1] | Eligibility | 254 |
| | | [2] | Duration | 255 |
| | | [3] | Rate Schedule | 255 |
| | | | [a] Rate Schedule | 255 |
| | | | [b] Tariff Degression | 255 |
| | | [4] | Cost-Sharing Scheme | 255 |
| | | [5] | Cap | 256 |
| | | [6] | Grid Connection and Mandatory Contracting Duty | 256 |
| | | [7] | Summary | 256 |
| | [B] | Comparison with German-Style FIT: Unique Schemes for PV | | 257 |
| §9.04 | Conclusions | | | 257 |

# Preface

Technology innovation and industrial revolution bring more convenient life and better life quality for the human society as a whole. However, water can overturn the boat as well as float it. Technology also leads to additional negative effects on human society, such as world war, environmental pollution, climate change, etc. The dominant approach in tackling the challenges resulted from technologies remains relying on technological solutions. We can find the application of this approach in dealing with environmental challenges, energy crisis, and climate change challenges, etc. For instance, even though the renewable electricity technology, such as hydropower, has been developed quite early, the recent prosperity of renewable electricity is highly related to the challenges of energy crisis in 1970s and the concerns of climate change since the discussion of United Nations Framework Convention on Climate Change (UNFCCC) in 1990s. How to use the law as a tool to facilitate the deployment the *renewable electricity* (RE) technology and possibly save the planet inspire the motivation of this book project.

Just like the problem in deployment of most novel technology, RE technology is relatively expensive and less cost-effective, compared to other traditional energy technology. The different forms of subsidies are usually provided to facility the RE technology from laboratory research and development (R&D), to demonstration and/or large-scale market application. According to study of the International Energy Agency (IEA)'s recommendation, multi-types of market deployment policy instrument are available to RE, including: bidding system, tax credit, obligation, tradable certificate, capital grants, government purchase, net metering, etc.[1] These subsidy scheme has played a role in contributing to the prosperity of RE in most of the countries since the 1970s.

Three 'key' market deployment schemes can also be identified from the experience in decades, they include: feed in tariff (FIT), tendering scheme, and RPS (renewable portfolio standard or Renewable obligation (RO)). The related literature

---

1. IEA renewable scheme http://s3.amazonaws.com/zanran_storage/www.iea.org/Content Pages/9895294.pdf at p. 85.

Preface

and research on policy, legal design and practical implementation, and the effectiveness of these three regimes has been widely studied and discussed in the western world.[2] Yet, relatively few 'legal' literature focuses on the development in Asia region, which also provide the rationale for this book.

Asian countries are definitely late comers in terms of RE technology, policy and legal regimes. Thus, similar to borrowing civil code, constitutional law, and criminal code from the western world to improve their existing traditional legal regime, in order to develop appropriate legal regime to promote RE, each country also tries to borrow the successful story of the western world in developing RE, such as: the successful FIT model from Germany, RPS model from the US. Therefore, to see how the RE legal regime in Asian region was affected by that of the western world, the idea of this book is to use the detailed design of FIT, RPS, tendering scheme, created by several important international level projects or database[3] as a parameter, and see how the main RE promotion regime in each country transforms these detailed design into their own RE legal regime. This book can be seen as a 'voluntary' research by extending European wide Res Legal in the jurisdiction of Asian region. Hopefully, it could contribute to further conversation or a EU-Asia forum between these two important RE regions.

The most recent influential factor to global RE policy and law is definitely the Fukushima accident of March 2011. This Fukushima issue has influenced the European RE policy and law, such as Germany, not to mention its huge impact on the Japan's neighbouring countries and itself. With a response to the energy or climate change policy after Fukushima accident, RE legal regime are also subject to likely reform. Two directions of reform can be identified: on the one hand, there is a group of countries, such as China and Thailand, seeking to modify their existing scheme by *fine-tuning* existing RE promotion scheme to reflect the need of post-Fukushima or climate change issues (Part II of this book). On the other hand, a group of countries adopting aggressive approach by changing track to other RE promotion scheme, such as: Japan and South Korea. (Part I of this book) After investigating into the detailed of country RE regime, this book will provide a cross-country-analysis on the RE legal regime in East Asia and Black Sea and Caspian Sea regions (Part III of this book).

After such an deep investigation into the RE legal regime and its implementation and the latest data on RE development, the preliminary finding is that even though most of the countries would 'formally' declare and emphasize the effects of successful RE promotion model of western society on their RE policy and legal regime, the 'substantial' legal context and detailed RE legal regime could tell another story. This 'promotion scheme' gap also impedes the development of RE in their jurisdiction. For

---

2. *See* e.g., Toby Couture et al., A Policymaker's Guide to Feed-in Tariff Policy Design, available at: http://www.energy.eu/publications/A_Policymakers_Guide_to_Feed-in_Tariffs_NREL.pdf; Clarisse Fräss-Ehrfeld, Renewable Energy Sources: A Chance to Combat Climate Change (2009); Miguel Mendonça et al., eds, Feed-in Tariffs: Accelerating the Deployment of Renewable Energy (2007); Res-Legal, Legal Sources on Renewable Energy, http://www.res-legal.eu/.
3. Res-Legal http://www.res-legal.eu/; IEA/IRENA, Renewable Energy Database. http://www.iea.org/policiesandmeasures/renewableenergy/.

instance, both the FIT in South Korea, China and Taiwan declares to be affected and inspired by the Germany FIT model. Yet, in terms of detailed design of cost-recovery issues, these countries either bizarrely introduced the idea of polluter-pays principle (comparing to the use-pays principle of Germany) or not fully passing all of the RE cost on the final consumers (comparing to the fully passing all of the cost of Germany model). Also, in terms of grid connection rules, most of Asian countries do not integrate the very comprehensive grid connection rules and expansion rules or favorable grid connection cost sharing scheme of Germany model. This kind of non-fully transposition of Germany model due to taking too much the political compromise or unknown of the essence of Germany Model may cast shadow on the further RE deployment.

What is the main reason behind this? It may be related to the dilemma between two different RE promotion directions. In Germany and many western countries, the RE policy and law focus on both the development of local RE *industry* and *real deployment*. The assumption is these two directions can complete each other. However, in most of Asian and manufacturing sector based countries, such as: Japan, South Korea, Taiwan, China, Thailand, etc., they are facing a dilemma in striking an appropriate balance between these two directions. Perhaps the real application may benefit the RE industry, but it may also have a potential impact on the electricity price, quality, and reliability, and hamper the existing manufacturing industry or energy intensive industry. Thus, there is always a critical debate between the supporter of the creation of new green jobs and preservation of existing and already-available grey jobs! Furthermore, for the less industrialized and developing countries, like Philippine and Indonesia, the lack of domestic RE industry may worsen their willingness in promoting RE's real deployment. Perhaps, there is similarity in selling *democracy* and RE *promotion scheme* from western world or international organizations, such as: IRENA. Certain Asian countries may be too poor to afford such luxury products of *democracy* and RE *technology*. However, the editors have to admit that there is also a situation of chicken first or egg first issue here!

Will this situation change in the near future? The Fukushima accident sent a *mix* message! For certain countries, particularly for those geologically located closer to the accident site and relatively richer countries, such as Japan, South Korea and Taiwan, the RE policy and legal regime become more active in promoting real deployment of RE. Yet, For other countries, particularly for those geologically located farer away from the accident site and relatively poor countries, like Indonesia and Philippines, the future of RE is blight. In general, it remains to be seen whether the further climate change talk and the worsening situation in Fukushima site would give some impetus to the political willingness or public supports to further RE development in the Asian region.

Finally, the publication of this book has to acknowledge the multi-funding support from National Science Council, National Tsing Hua University and Ministry of Education (NSC 102-3113-P-007-002-; NSC101-2410-H-007-024-MY2; Excellent Centre Project of NTHU and Ministry of Education: A Study on Low-carbon Policy, Economics, Law at a Post-Kyoto New Situation: Focus on Carbon Market and the Legal Roadmap for Low Carbon Technology Development.) Also, the draft of the articles have been presented and discussed in the International Joint Conference on Changing

Energy Law and Policy in Asia Region on 17–18 October 2013 at the venue of 9F, TSMC Building, the campus of National Tsing Hua University, Taiwan. The conference funding is provided by Center for Energy and Environmental Research, Research Center for Humanities and Social Science, Bioethics and Law Center, and Office of Research and Development of NTHU, and National Science council. The efforts of working team are very much appreciated as well. Hopefully, in the short term, the conference can become an important annual energy law research and publication platform in the Asian region; in the long term, this conference can be expanded to become an Euro-Asia energy law forum.

Anton Ming-Zhi Gao
Chien Te Fan

*At 8F., TSMC Building, Institute of Law for Science & Technology,*
*National Tsing Hua University, Taiwan*

# Acknowledgement

The editors would like to acknowledge the support funding from National Science Council of Taiwan (Project number: NSC101-2410-H-007-024-MY2;102-3113-P-007-002) and Ministry of Education (Project title: A Study on Low-carbon Policy, Economics, and Law at a Post-Kyoto New Situation, Excellent Centre Project of NTHU).

PART I New Renewable Electricity Promotion Regime after the Fukushima Accident

CHAPTER 1
# Renewable Energy-Related Policies and Institutions in Japan: Before and after the Fukushima Nuclear Accident and the Feed-In Tariff Introduction

*Kanako Morita & Ken'ichi Matsumoto*[*]

## §1.01 INTRODUCTION

Exploring how to enhance the use of renewable energy is a key national political agenda in Japan aimed at achieving national energy security and contributing to solving climate change problems as stated in the 1997 Act on the Promotion of New Energy Usage. This chapter focuses on the three major renewable energy-related schemes in Japan: the Renewable Portfolio Standard (RPS) Scheme (the Act on Special Measures Concerning New Energy Use by Operators of Electricity Companies), the Net-metering Scheme for Photovoltaic Power (the Act on the Promotion of the Use of Non-fossil Energy Sources and Effective Use of Fossil Energy Materials by Energy Suppliers or the Excess Electricity Purchasing Scheme for Photovoltaic Power), and the Feed-in Tariff (FIT) Scheme for renewable energy, which began in July 2012.

In Japan, the term 'new energy', which means an energy alternative to petroleum, has been commonly used in writing energy-related laws and policies.[1] However,

---

[*] This research was supported by grants from the Global Environment Research Fund of the Ministry of the Environment of Japan S-11 and JSPS KAKENHI 24710046. We would also like to thank Mr. Takashi Kitamura, Agency for Natural Resources and Energy, Ministry of Economy, Trade and Industry, and Mr. Shuta Mano and Mr. Keiji Kimura at the Japan Renewable Energy Foundation, for the useful interviews and providing us information on renewable energy policies in Japan.
1. New energy is defined as photovoltaic power, wind power, mid- and small-scale hydropower, geothermal power, solar heat utilization, temperature-difference heat utilization, biomass power,

in this chapter, we basically use the term 'renewable energy', the major sources of which overlap with those of new energy, to focus more on the aspects of environmental integrity and climate-change mitigation, and to include energy such as hydropower and geothermal power, which are not covered by new energy.

The Japanese energy self-sufficiency rate was only 4.8% in 2010, including the use of hydro, geothermal, solar, and biomass power.[2,3] Like other major industrialized countries, Japan is highly dependent on fossil fuels, which account for more than 80% of energy supply. However, because much fossil fuel is imported from the politically unstable Middle East, and because the energy demands of emerging countries such as China and India are increasing, and these countries are trying to secure their energy supply, it is becoming more difficult for Japan to rely on importing most of its energy as fossil fuel. Therefore, Japan needs to develop its own energy sources.

One of the energy sources that Japan has focused on and expanded in order to depart from fossil fuel dependence has been nuclear power. The Act on Promotion of Global Warming Countermeasures was adopted in 1998, and nuclear power has gained attention as one of the important energy sources that reduce greenhouse gas (GHG) emissions. In February 2011, there were 54 commercial nuclear power plants operating in Japan. However, the Great East Japan Earthquake, followed by the Fukushima nuclear accident on 11 March 2011, highlighted the safety issues of using nuclear power. Currently, as of June 2013, most nuclear power plants are offline; only two are currently in operation.[4]

The Democratic Party of Japan has contributed to increasing renewable electricity by providing policy targets on renewable energy, such as the bill of Basic Act on Global Warming Countermeasures which included a target of raising the share of renewable energy out of the total primary energy supply to 10% by 2020, and the decision of the Energy and Environmental Council of the Japanese government made in September 2012 which aim for zero nuclear power generation by the 2030s. However, these were scrapped after the dissolution of the lower house in November, 2012.

As an alternative energy, renewable energy will be one of the most important elements in securing Japan's national energy supply and solving climate change problems. Although multiple national renewable energy-related policies and institutions were introduced to diffuse the generation of renewable energy in Japan after the oil crisis in the 1970s, the current share of renewable energy in Japan in 2011 (excluding hydropower) is only 1.4% of total power generation.[5]

---

biomass heat utilization, biomass fuel fabrication, and snow and ice heat utilization. Renewable energy includes large-scale hydropower and ocean energy in addition to new energy (Agency for Natural Resources and Energy (ANRE), *What is new energy*, http://www.enecho.meti.go.jp/energy/newenergy/new/p1.html [accessed 28 Jun. 2013, in Japanese]).
2. Nuclear power is not considered domestic energy.
3. Ministry of Economy, Trade and Industry (METI) Energy White Paper 2012. Energy Forum: Tokyo (in Japanese) (2012).
4. Japan Nuclear Technology Institute, *Operation of nuclear power plant*, http://www.gengikyo.jp/db/fm/plantstatus.php?x=d (accessed 28 Jun. 2013; in Japanese).
5. ANRE, *About renewable energy*, http://www.enecho.meti.go.jp/saiene/data/kaitori/2012kaitori.pdf (accessed 28 Jun. 2013; in Japanese).

This chapter outlines previous and current national renewable energy-related policies and institutions in Japan, and describes their effects and challenges. Section §1.02 describes the previous renewable energy-related schemes, with a focus on two major schemes, the RPS Scheme and the Net-metering Scheme for Photovoltaic Power. The section discusses the effects of the schemes, and the reasons why the share of renewable energy has not expanded dramatically. Section §1.03 explains the FIT Scheme for renewable energy, which began in July 2012, and describes the background to the scheme and its expected effects. Section §1.04 summarizes the chapter and discusses effective renewable energy policies.

## §1.02 A BRIEF OVERVIEW OF RENEWABLE ENERGY-RELATED POLICIES AND INSTITUTIONS IN JAPAN BEFORE THE INTRODUCTION OF THE FIT SCHEME IN 2012

### [A] Renewable Energy-Related Policies and Legal Measures

During the period of rapid economic growth up to the 1970s in Japan, energy consumption there greatly increased. For example, the total final energy consumption more than doubled from fiscal year FY 1965 to FY 1970.[6] However, in the 1970s, Japan suffered large economic impacts from the oil crises (first in 1973 and then in 1979), and the importance of alternative energy received greater recognition. Japan began to address renewable energy in earnest.

This section gives an overview of renewable energy-related policies and institutions in the period from the oil crisis in the 1970s to 2012. The first major policy for renewable energy was the Sunshine Project[7] established in 1974. This commenced under the direction of the Agency of Industrial Science and Technology in Japan. The project aims to enhance research and development (R&D) on oil-alternative energy technologies with a focus on solar, geothermal, coal, and hydrogen energy, in order to supply sufficient energy to meet Japanese energy demands in the following decades.[8] Afterward, the Act on the Rational Use of Energy (1979) and the Act on the Promotion of Development and Introduction of Alternative Energy (1980) were adopted.

In 1997, the Act on the Promotion of New Energy Usage came into force to enhance the use of so-called new energy. This act specifies the role of each actor, including the country, local government, business operators, and citizens, and provides financial support measures for business operators who use new energy.[9]

The RPS Act was promulgated in 2002 and enacted in 2003. It aims to further the use of new energy by requiring certain electricity companies to use a certain amount of electricity generated from new energy annually. Compared with previous policies and

---

6. Institute of Energy Economics, Japan (IEEJ), Handbook of Energy and Economic Statistics in Japan 2013. The Energy Conservation Center, Tokyo (2013).
7. In 1993, the Moonlight Project (an energy-saving-technology research and development project established in 1978) was integrated into the Sunshine Project, and restarted as the New Sunshine Project.
8. METI, *supra* n. 3, at 189.
9. *Ibid.*

institutions, the RPS scheme is a more direct policy to enhance renewable energy. Types of energy covered in the act are photovoltaic (PV), wind, biomass, medium- and small-scale hydro, and geothermal power. The annual targets for new-energy electricity use by electricity retailers are established by the Ministry of Economy, Trade and Industry (METI). For example, in 2007, the target for 2014 was set as 16.0 TWh.[10]

In 2009, the Net-metering Scheme for Photovoltaic Power was launched. Under the scheme, electricity companies are mandated to purchase any surplus electricity that has been generated by customers' PV facilities. The cost of buying back surplus electricity is borne by electricity customers (residential and non-residential) in the form of a PV promotion surcharge.[11] This scheme has evolved into the FIT Scheme launched in 2012.

In parallel with the development of energy-related policies and institutions, there has been a growing interest in environmental integrity and climate change problems. As described above, in 1998, a year after the Act on the Promotion of New Energy Usage came into force, the Act on Promotion of Global Warming Countermeasures was adopted; the act calls for the national government to implement the Kyoto Protocol Target Achievement plan (where the target for Japan was a 6% reduction in GHG emissions compared with the base year 1990 over 2008–2012). On 12 March 2010, the bill of the Basic Act on Global Warming Countermeasures was approved by Prime Minister Yukio Hatoyama and his cabinet and submitted to the Diet. The bill included mid- and long-term GHG emission reduction targets, which are a reduction to 25% below the 1990 level by 2020, and a reduction to 80% below the 1990 level by 2050. The bill also included a target, raising the share of renewable energy to 10% of the total primary energy supply by 2020. Although this bill was scrapped after the dissolution of the lower house in 16 November 2012, the discussion on the bill contributed to raising public attention on renewable energy promotion to some degree. Furthermore, before the Fukushima Nuclear accident in 2011, one of the key forms of energy that could be used to address climate change was nuclear power, but since the accident, there has been a call to use renewable energy that has environmental integrity.

In addition to the above major obligations to enhance renewable energy, there are other schemes such as tax reductions (e.g., green investment tax credits), subsidies (e.g., for the cost of implementing measures to support the use of renewable energy heat, and installation of residential PV systems), and loans (e.g., for implementing environmental and energy measures).[12] Additionally, there are several voluntary schemes designed to enhance the use of renewable energy, such as the Green Power Certificate, which is a system that enhances the use of renewable energy by the introduction of trading certificates based on environmental added value from the

---

10. ANRE, *RPS Act website*, http://www.rps.go.jp/RPS/new-contents/top/toplink-english.html (accessed 28 Jun. 2013).
11. METI, *Approval of electricity charges followed by the fixing of FY 2012 Photovoltaic Power Promotion Surcharge Rates*, http://www.meti.go.jp/english/press/2012/0125_02.html (accessed 28 Jun. 2013). As of June 2013, the target for the period after FY 2014 has been changed to zero.
12. ANRE, *Support system for renewable energy introduction*, http://www.enecho.meti.go.jp/saiene/support/ (accessed 28 Jun. 2013; in Japanese).

electrical power generated using, for example, wind, PV, hydro, biomass, and geothermal power.[13]

### [B] Implementation and Effects of Major Renewable Energy-Related Schemes

As described in the previous section, Japan has introduced multiple policies and institutions to enhance the generation of renewable energy, but with only limited effect. Figure 1.1 shows the share of energy sources in the national primary energy supply in Japan. It shows that renewable energy (excluding hydropower) accounted for only 3.7% of the total primary energy supply in 2010.[14] summarized the key barriers to increasing generation of renewable energy in Japan as the geographical restrictions peculiar to Japan (e.g., limited space suitable for large-scale facilities), high costs (the costs of the development and introduction of renewable energy are very high for companies or individuals at present), and instability in its supply (wind and PV power are both subject to natural conditions and the variability of these resources makes the electricity sources unstable).

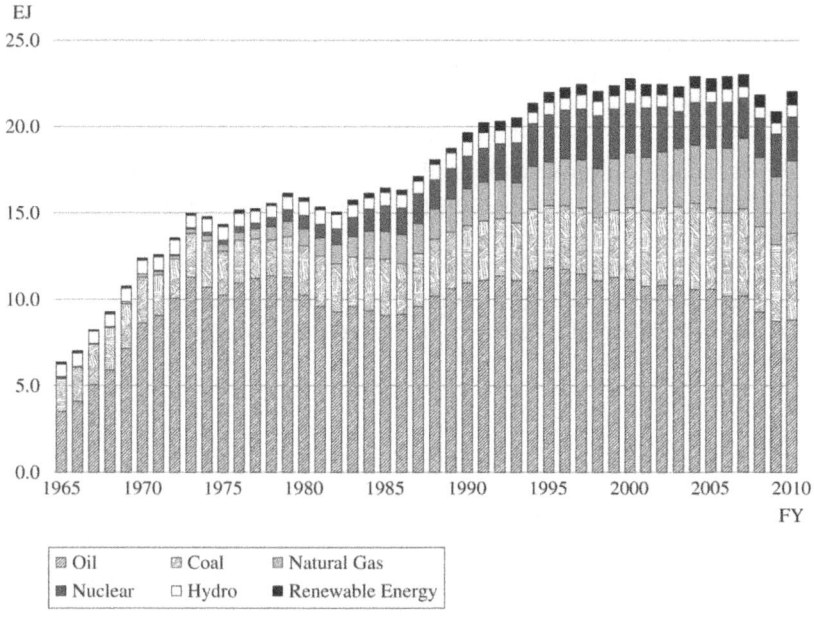

Figure 1.1   National Primary Energy Supply in Japan[15]

---

13. Ministry of the Environment (MOE), *What is a green electricity certificate?*, http://www.env.go.jp/earth/ondanka/greenenergy/ (accessed 28 Jun. 2013; in Japanese).
14. Meritas *On the horizon – renewable energy in Asia*. Meritas Report (2012).
15. METI, *supra* n. 3, 97.

The following describes the implementation and effects of the two major renewable energy-related policies in Japan prior to 2012, the RPS Scheme and the Net-metering Scheme for Photovoltaic Power.

In the case of the RPS Scheme, the amount of new energy used by electricity companies was increased by setting an obligatory amount of new-energy use. Annual targets were set for the use of new-energy electricity by electric retailers for eight years. For example, in 2007, the annual targets were 8.7 TWh/year in FY 2007, and 16 TWh/year in FY 2014.[16]

In FY 2010, the RPS Scheme required 53 electricity companies (10 major electricity-generating companies, five specified electricity companies, and 38 specified-scale electricity companies) to use a total of 11,015 GWh of renewable energy-sourced electricity (renewable electricity).[17] In FY 2010, the total amount of renewable electricity supplied to electricity companies by renewable electricity facilities was 10,246 GWh (8,873 GWh in FY 2009).[18] There is a banking system such that if an electricity company supplies renewable electricity in excess of the required amount for that fiscal year, it may carry over the excess amount to the next fiscal year.[19] From FY 2009 to FY 2010, 6,406 GWh was carried over.

Among the forms of renewable energy, PV power is one into which Japan has put a great deal of effort. From 1994 to 2005, and from 2008 to the present, there have been installment subsidies for residential PV power generation. In 2009, the Net-metering Scheme for Photovoltaic Power, which sets the purchasing price of PV power that electricity companies are obliged to purchase, was launched. The purchase prices from April to June 2012 were set at JPY 42/kW (double power generation JPY 34/kWh) for residential facilities less than 10 kW, JPY 40/kWh (double power generation JPY 32/kWh) for non-residential facilities and residential facilities of 10 kW or more, and JPY 24/kWh (double power generation JPY 20/kWh) for facilities installed before FY 2010.[20] In July 2012, the scheme became part of the FIT Scheme, which is explained in section §1.03.

The share of residential PV power generation has increased through the introduction of the scheme, although it is difficult to distinguish the effects of separate schemes. The installation of PV systems increased from 2,627 MW in FY 2009 to 4,910 MW in FY 2011.[21]

---

16. ANRE, *RPS Act website*, http://www.rps.go.jp/RPS/new-contents/top/toplink-1.html/ (accessed 29 Jun. 2013; in Japanese).
17. METI, *Report on the implementation of the RPS Act in FY2010*, http://www.meti.go.jp/english/press/2011/0714_01.html (accessed 29 Jun. 2013).
18. *Ibid.*
19. *Ibid.*
20. METI, *Purchasing prices for April to June 2012 determined; Excess Electricity Purchasing Scheme for Photovoltaic Electricity*, http://www.meti.go.jp/english/press/2012/0301_02.html (accessed 29 Jun. 2013).
21. Institute of Energy Economics, Japan (IEEJ), Handbook of Energy and Economic Statistics in Japan 2013. The Energy Conservation Center, Tokyo (2013).

## [C] Challenges Facing Major Renewable Energy-Related Schemes

The previous section showed that the RPS Scheme and the Net-metering Scheme for Photovoltaic Power had some effects in increasing the use of renewable electricity. Although it is not clear how much specific schemes directly contributed to the increase in use of renewable electricity since there were multiple renewable energy-related schemes, Figure 1.2 shows that the total generation of renewable electricity has increased gradually since 2003. In particular, there has been a large increase in PV power generation since the Net-metering Scheme for Photovoltaic Power was introduced.

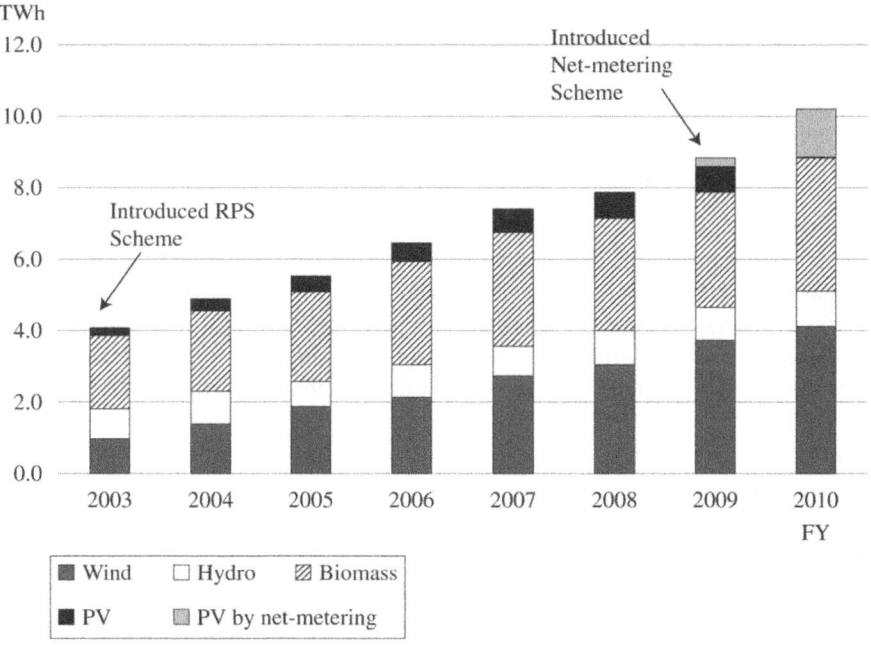

*Figure 1.2   Renewable Electricity Production[22]*

However, as Figure 1.1 illustrated, the share of renewable energy is still small compared with the total primary energy supply in Japan. This indicates that the multiple renewable energy-related schemes including the RPS Scheme and the Net-metering Scheme for Photovoltaic Power were not successful in dramatically increasing the generation of renewable electricity.

The RPS Scheme was established with perceived multiple strengths, such as (1) there being less risk in introducing renewable electricity because the scheme sets an

---

22. ANRE, *About the FIT Scheme for renewable energy* http://www.hkd.meti.go.jp/hokne/saiene_nintei/seido.pdf (accessed 5 May 2013; in Japanese).

obligatory amount of renewable energy that is equitable among electricity business operators; (2) the scheme being equitable in cost burden and in market competition; and (3) the scheme having cost-effectiveness through market competition among electricity business operators. However, the RPS Scheme also has the limitation that electricity companies are obliged to purchase a certain amount of electricity. Oshima (2012)[23] explains that there are three main problems in employing a quota system in Japan: (1) the low level of the obligatory amount (in FY 2011, the annual targets of use of electricity derived from new energy by electricity retailers only accounted for 1.25% of the total generation of renewable electricity in that year), (2) inclusion of waste power generation as biomass of renewable energy (the cost of which is cheaper than that of renewable energy), and (3) the high likelihood of supporting existing equipment that does not need support, rather than supporting new equipment needed to generate and use renewable electricity. There are also lesser problems such as (4) the exclusion of a large part of geothermal power and small-scale hydropower larger than 1,000 kW, and (5) the setting of only the maximum purchasing price and the short period of purchasing goals and the introduction of the banking system that allows the electricity company to transfer an excess to the next fiscal year, meaning that there is a lack of incentive for electricity retailers to use more than the obligatory amount of renewable energy.[24]

The Net-metering Scheme for Photovoltaic Power is a new scheme restricting the purchasing price, and has better promoted the use of PV facilities than the RPS Scheme. Although this scheme has contributed to increasing PV power generation, the challenges are that it does not support other forms of renewable electricity such as hydro and wind power, and that excess electricity is purchased only from residential and non-residential PV installments and not from power generators. In 2012, this scheme evolved into the FIT scheme, which is explained in the next section.

## §1.03 FIT SCHEME IN JAPAN

The FIT Scheme in Japan (the Act on Special Measures Concerning the Procurement of Renewable Energy by Electricity Utilities) obliges electricity companies to accept applications of electricity supply contracts requested by renewable electricity producers at a fixed purchase price for a long-term period guaranteed by the Japanese government. The scheme regulates the procedures followed by the government in deciding purchase prices and periods, the certification of facilities, the collection and adjustment of surcharges related to purchase costs, and terms by which companies can reject the contracts.

The purpose of the FIT Scheme is to promote the use of renewable electricity, in addition to enhancing international competitiveness, industrial development, local revitalization, and economic development (Article 1 of the FIT Scheme). The scheme is

---

23. Kenichi Oshima, *Political Economics of Renewable Energy*. Toyo Keizai: Tokyo (in Japanese) (2012).
24. Institute for Sustainable Energy Policies (ISEP), Renewables Japan Status Report 2013. Nanatsumori Shokan: Tokyo (in Japanese) (2013).

expected to reduce uncertainty in investment recovery for renewable electricity facilities, and encourage investment to increase the generation of renewable electricity (and renewable energy itself).[25]

### [A] Reasons for Introducing the FIT Scheme

Before the FIT Scheme was enforced, as mentioned above, the RPS Scheme obliging electricity companies to purchase a certain 'amount' of electricity rather than to purchase electricity at a certain 'price' steadily increased the generation of renewable electricity. At that time, the Japanese government left some options for energy sources to the market since such a pricing policy might distort markets. However, the government did introduce an experimental pricing scheme, the Net-metering Scheme for Photovoltaic Power, in 2009, which was very successful in promoting the use of PV facilities. However, after the Great East Japan Earthquake in March 2011, the situation relating to energy has largely changed. Following the Fukushima nuclear accident, a large part of the Japanese population expressed a wish to abandon nuclear power generation, which lacks security, and to increase the generation of renewable energy. Prime Minister Naoto Kan and his cabinet decided that the share of renewable energy needs to be expanded even if the markets are distorted. Consequently, a more ambitious scheme, the FIT Scheme, was introduced in 2012.

The FIT Scheme in Japan had been under consideration since 2009.[26] The Japanese government had already been planning the FIT Scheme before the Fukushima Nuclear accident, but the accident accelerated the implementation of the scheme.

### [B] Architecture of the FIT Scheme

#### *[1] Overview of the Process of the FIT Scheme*

Figure 1.3 is an overview of the FIT Scheme in Japan. Renewable electricity producers (including households) need their power-generating facilities to be certified by the METI.[27] Each electricity company purchases renewable electricity generated by the certified facilities at a fixed price for a period guaranteed by the government. The purchase costs of renewable electricity are shared by all consumers who purchase electricity from electricity companies in proportion to the volume of electricity they

---

25. Takashi Kitamura, *Situation of the FIT scheme and challenge toward substantial expansion of renewable energy.* Energy and Resources 34 (3), 129–133 (in Japanese) (2013).
26. Agency for Natural Resources and Energy (ANRE), *Feed-in tariff scheme for renewable energy*, http://www.meti.go.jp/english/policy/energy_environment/renewable/pdf/summary201209.pdf (accessed 5 May 2013).
27. PV facilities whose capacity is less than 50 kW are certified through the Japan Photovoltaic Energy Association, and others are certified through the Bureau of Economy, Trade, and Industry of each area. Finally, the METI certifies the facilities. When specifications of a certified facility change, the facility needs to be certified again. In the case that the electricity supply by extension or repowering is measureable and can be confirmed, the electricity can be purchased (ANRE, 2012a). ANRE, *About the FIT Scheme for renewable energy* http://www.hkd.meti.go.jp/hokne/saiene_nintei/seido.pdf (accessed 5 May 2013; in Japanese).

have used.[28] The Surcharge Adjustment Organization[29] collects, calculates, and distributes the surcharge (Articles 8-18 of the FIT Scheme). Renewable electricity producers need to have two contracts with electricity companies, a specific contract and a grid connection contract (Articles 4 and 5 of the FIT Scheme). The former is a contract for the purchase and sale of generated electricity, and the latter is a contract for the connection to grids of electricity companies.

*Figure 1.3   Overview of the FIT Scheme in Japan[30]*

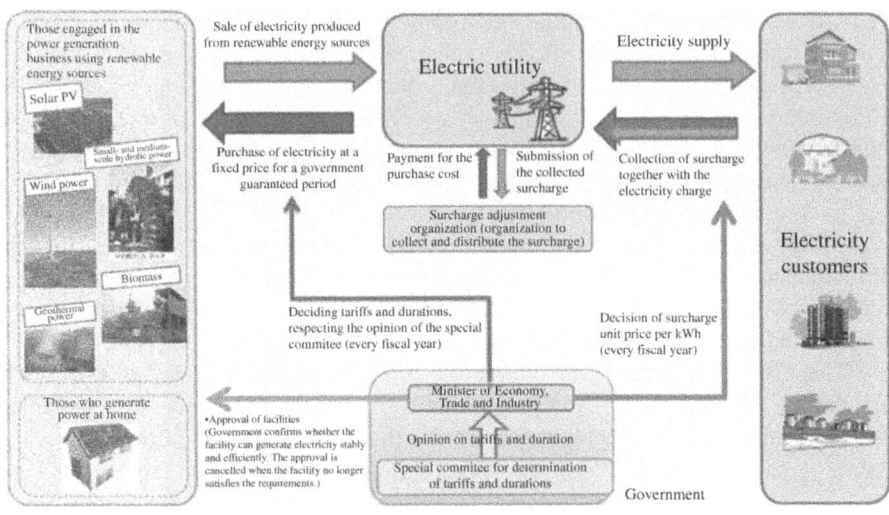

### [2]   Purchase Prices and Periods under the FIT Scheme

The FIT Scheme covers PV, wind, hydro (small- and medium- scale), geothermal, and biomass power (Article 2; Table 1.1).[31] However, not all facilities for each type of renewable energy are included in this scheme. Table 1.1 summarizes the purchase prices and periods for each form of renewable electricity under the FIT Scheme. The prices and periods are set according to the classification, installation mode, and scale of renewable electricity facilities (Article 3 of the FIT Scheme). The purchase prices and periods of PV power in FY 2012 are more favorable for renewable electricity producers

---

28. The surcharge is determined in units of JPY/kWh. The surcharge for the promotion of PV power under the Net-metering Scheme for Photovoltaic Power starting in FY 2009 is also imposed on electricity use (until September 2014).
29. The role of the Surcharge Adjustment Organization is described in Arts 19-30 of the FIT Scheme.
30. Agency for Natural Resources and Energy (ANRE), *Feed-in tariff scheme for renewable energy*, http://www.meti.go.jp/english/policy/energy_environment/renewable/pdf/summary201209.pdf (accessed 5 May 2013).
31. Other types of renewable energy that are now in the R&D phase will be included in the scheme in the future, e.g., renewable energy is expected to be diffused to some extent (Art. 2.4.6 of the FIT Scheme).

than those in the previous Net-metering Scheme for Photovoltaic Power. Basically, the full amount of electricity generated is purchased at a fixed price in a fixed period. For residential PV power generation, however, only surplus electricity is purchased (i.e., net-metering[32]). This provides an incentive for households to save energy and to stress the continuity of the policy (i.e., the Net-metering Scheme for Photovoltaic Power). The price to be adopted is that when the electricity company receives the application form for a contract of access to the electric power system or when the METI certifies the facility, whichever is later. The period commences at the time of initiation of electricity supply under a specific contract.

*Table 1.1    Classifications of Renewable Electricity, and Purchase Prices and Periods under the FIT Scheme in Japan*

|  |  | Price (Including Tax; JPY/kWh) | | Period (year) | Pre-tax IRR (%) |
|---|---|---|---|---|---|
|  |  | 2012 | 2013 |  |  |
| PV* | ≥10 kW | 42 | 37.8 | 20 | 6 |
|  | <10 kW | 42 | 38 | 10 | 3.2 |
|  | <10 kW (with private power facilities) | 34 | 31 | 10 | – |
| Wind | ≥20 kW | 23.1 | | 20 | 8 |
|  | <20 kW | 57.75 | | | 1.8 |
| Hydro | 1,000–30,000 kW | 25.2 | | 20 | 7 |
|  | 200-1,000 kW | 30.45 | | | 7 |
|  | <200 kW | 35.7 | | | 7 |
| Geothermal | ≥15,000 kW | 27.3 | | 15 | 13 |
|  | <15,000 kW | 42 | | | 13 |
| Biomass** | Biogas | 40.95 | | 20 | 1 |
|  | Wood-fired power plant (timber from forest thinning) | 33.6 | | | 8 |
|  | Wood-fired power plant (other woody materials) | 25.2 | | | 4 |
|  | Waste | 17.85 | | | 4 |
|  | Wood-fired power plant (recycled wood) | 13.65 | | | 4 |

\* Biomass that does not affect existing applications, such as paper and pulp, are available.
\*\* In the case that systems of multiple types of renewable energy are installed, if the electricity supply from each facility is measurable and confirmed, each facility is certified and an

---

32. ANRE, *Major terms regulated in the FIT Scheme that should be assumed in the discussions of the Procurement Price Calculation Committee* http://www.meti.go.jp/committee/chotatsu_kakaku/pdf/011_s01_00.pdf (accessed 5 May 2013; in Japanese).

appropriate purchase price is adopted for each. If only the total amount is measurable, the price adopted for the lower-purchase price facility is adopted.

Source: ANRE (2011, 2012a) and ANRE, *Purchase price and periods*, http://www.enecho.meti.go.jp/saiene/kaitori/kakaku.html (accessed 5 May 2013; in Japanese).

In the Net-metering Scheme for Photovoltaic Power, the price was set by the METI. However, in the FIT Scheme, the purchase prices, periods, and classifications are discussed and reviewed every fiscal year by the Procurement Price Calculation Committee (Articles 31–37 of the FIT Scheme),[33] and the Minister of the METI makes the final decisions taking into consideration the discussions before the next fiscal year starts (Article 3 of the FIT Scheme). The purchase prices, periods, and classifications are determined so that the surcharge paid by consumers of electricity is not too high. The prices are determined according to costs usually incurred in generating electricity efficiently and the average expected power generation per facility, considering the present situation of the introduction of renewable electricity, a reasonable profit for renewable electricity producers, and actual costs in previous examples (i.e., information such as prices of materials of a power generation system obligatorily reported by the certified renewable electricity producers;[34] Article 3 of the FIT Scheme). In the first three years, however, the prices are set to be high enough for the producers to make a profit in order to activate renewable energy markets (Article 7 of Additional Clause of the FIT Scheme). The purchasing periods are determined considering basic periods between the start of electricity supply and the update of important components of the facilities (Article 3 of the FIT Scheme).

Table 1.1 shows that only the price of PV power has changed from FY 2012 to FY 2013. This is because there are too few newly-operating facilities for other forms of renewable electricity, and thus, there is insufficient evidence with which to update prices.[35]

### [3]   Certification of Producers of Renewable Electricity

Renewable electricity producers are requested to obtain certification of their power generation facilities if they are to be involved in the FIT Scheme. The basic standards that needs to be met for certification are (1) the facility must be able to stably and efficiently generate electricity during the purchase period; (2) the facility must be able to transparently and fairly measure the amount of renewable electricity that is supplied to the electricity company; and (3) the facility to be used for power generation must be

---

33. The amount of surcharge (*see* section §1.03[B][5]) is considered when the Procurement Price Calculation Committee discusses prices, periods, and classifications.
34. For FY 2012, the first year of the FIT Scheme, the prices were determined by interviewing producer associations and each producer on the costs of power generation.
35. Procurement Price Calculation Committee, *Opinions on purchasing prices and periods in FY 2013*, http://www.meti.go.jp/committee/chotatsu_kakaku/pdf/report_002_01_00.pdf (Accessed 6 May 2013; in Japanese).

specified in detail (Article 6 of the FIT Scheme).[36] When there is any change to an already certified facility, it is necessary to obtain certification for the facility anew. When the incremental amount of electricity supplied through additional installation or repowering can be clearly measured and this fact can be confirmed by a wiring diagram or the like, the incremental output can be made subject to purchase.

### [4]   Purchase of Electricity by Electricity Companies

Since the amount of electricity that electricity companies have to purchase is not stipulated, they are basically obliged to purchase the full amount of electricity generated by the certified facilities (Articles 4 and 5 of the FIT Scheme). On the other hand, there are some conditions under which the electricity companies can refuse the purchase and connection of electricity generated (Article 5 of the FIT Scheme and Articles 4–6 of the Ordinance of METI). These are, for example, (1) when renewable electricity producers do not bear the necessary costs for the connection stipulated in the Ordinance of METI;[37] (2) when the connection has a significant adverse effect on the securement and supply of electricity by electricity companies; (3) when it is reasonably expected that the connection will exceed the capacity of transmission at the point where the producers wish to connect, and (a) the electricity companies provide reasonable evidence in writing to the producers and (b) they indicate the connecting points that are economically reasonable for the producers in writing to them; and (4) it is expected that electricity companies will receive an electricity supply from renewable electricity producers exceeding their capacity, even if they control the power without any compensation within the limits of 30 days in a year.[38] In any case, the electricity companies are requested to provide reasonable evidence in writing to the producers. Each electricity company controls power fluctuation due to fluctuation of renewable electricity by thermal power and hydropower generation. The capacity to control the fluctuation is determined by each company in each area and also on the local scale in the area. The capacity is relatively high in areas with large electricity demand (e.g., Kanto and Kansai areas).

### [5]   Surcharge Adjustment under the FIT Scheme

Since the amount of renewable electricity produced is different among different areas (e.g., the amount is largest in the Hokkaido area), the Surcharge Adjustment Organization, consigned to the Green Investment Promotion Organization, collects, calculates, and distributes the surcharge. First, each electricity company submits the

---

36. The details of the standards are determined by the renewable energy source (Art. 6 of the FIT Scheme and ANRE, 2011).
37. Renewable electricity producers bear the cost for power wires, voltage regulators, electricity meters, etc. (Art. 5 of the Ordinance of METI). Therefore, the deep method is basically adopted for them (*see also* Art. 6 of the Ordinance of METI).
38. Details are given in Art. 5 of the FIT Scheme and Arts 4–6 of the FIT Ordinance (the Enforcement Regulations).

surcharge collected from electricity consumers to the Surcharge Adjustment Organization.[39] Two groups of consumers are partially or fully exempted from the surcharge: (1) those affected by the Great East Japan Earthquake (exempted from surcharge payment from August 2012 to April 2013; Article 9 of the additional clause of the FIT Scheme) and (2) enterprises that consume a large amount of energy (exempted from payment of 80% or more of the surcharge; Article 17 of the FIT Scheme).[40] The organization then calculates the payment amount for the purchase costs of each electricity company, and distributes it. The Minister of METI determines the amount of surcharge every fiscal year[41] (Article 12 of the FIT Scheme). In FY 2012, the surcharge was JPY 0.22/kWh and was updated to JPY 0.35/kWh in FY 2013.

*[6]   Revision of the FIT Scheme*

The scheme is revised when necessary by the legislative branch of the government. The procedure followed in revising the scheme is stipulated as (1) when the Basic Energy Plan is revised, the scheme will also be revised if necessary; (2) the scheme shall be examined at least once every three years and necessary measures shall be implemented; and (3) the scheme shall be revised drastically taking into consideration the enforcement of the scheme by 2020 (Article 10 of Additional Clause of the FIT Scheme).

### [C]   Implementation and Effects of the FIT Scheme in Japan

*[1]   Production of Renewable Electricity under the FIT Scheme*

Table 1.2 lists renewable electricity facilities introduced between April 2012 (the start of the FIT Scheme) and March 2013.[42] The table shows that the introduction of renewable electricity was greatly promoted after the FIT Scheme was introduced. Renewable electricity producers tended to introduce PV facilities between 10 kW and 50 kW. This is because only surplus electricity is purchased, the purchase period is shorter for facilities less than 10 kW, and legal restrictions are stricter for facilities greater than or equal to 50 kW.[43] At the prefectural scale, PV facilities less than 10 kW

---

39. Electricity companies are permitted to collect a surcharge that exceeds their power generation costs.
40. A reduction in the surcharge of 80% or more is to be provided to business facilities whose annual electricity use exceeds an amount to be set forth in the Implementing Regulations, upon application by a business operator whose ratio of electricity use (in kWh) to sales volume (per thousand yen) (a) exceeds 8 times the average ratio in the manufacturing industry (if a manufacturer) or (b) exceeds the average ratio in the non-manufacturing industry (if a non-manufacturer) by a factor to be determined in the Enforcement Regulations.
41. surcharge = (the amount of total purchase − avoided expenses + administrative fee of surcharge adjustment organization) / annual power supply by electricity companies.
42. ANRE, *Status of introduction of renewable electricity facilities*, http://www.enecho.meti.go.jp/saiene/kaitori/dl/setsubi/201305setsubi.pdf (accessed 2 Sep. 2013; in Japanese). The information is updated almost every month.
43. ANRE, *Basic guidelines for PV field test* http://www.enecho.meti.go.jp/energy/newenergy/basic_2011.pdf (accessed 26 May 2013; in Japanese).

are installed in urban areas while PV facilities greater than or equal to 10 kW are installed throughout the country.[44] The output in Hokkaido, where multiple solar power mega-plants are planned, is the most distinctive—PV facilities of 2,059 MW are certified.[45] In addition, PV facilities are popular because environmental assessment is not required and the installation is easier than the installation of facilities for other forms of renewable electricity.[46]

Table 1.2 Introduction of Renewable Energy Facilities in Japan[47]

| Facilities for Renewable Electricity (Type of Source) | Cumulative Capacity of Facilities as of the End of FY 2011 | Combined Total Capacity of Facilities that Started Operation in FY 2012 | Facilities Certified until the End of FY 2012 |
|---|---|---|---|
| PV (for households) | 4,400 MW | 1,269 MW (300 MW from April to June) | 1,342 MW |
| PV (other than households) | 900 MW | 706 MW (2 MW from April to June) | 18,681 MW |
| Wind | 2,600 MW | 363 MW (zero from April to June) | 798 MW |
| Small and medium hydro (1,000 kW or more) | 9,400 MW | 1 MW (1 MW from April to June) | 61 MW |
| Small and medium hydro (less than 1,000 kW) | 200 MW | 23 MW (1 MW from April to June) | 10 MW |
| Biomass | 2,300 MW | 36 MW (6 MW from April to June) | 194 MW |
| Geothermal | 500 MW | 1 MW (zero from April to June) | 4 MW |
| Total | 20,000 MW | 2,079 MW | 21,090 MW |

Many facilities operating under the RPS Scheme became part of the FIT Scheme (Article 12 of Additional Clause of the FIT Scheme). Under the RPS Scheme, facilities of 5,287 MW generated 9,360 GWh of electricity. About 70% (3,722 MW or 6,650 GWh) of output moved to the FIT Scheme,[48] including most wind power facilities (2,522 MW

---

44. ANRE, *Status of certification of renewable electricity facilities*, http://www.enecho.meti.go.jp/saiene/kaitori/index.html (accessed 26 Jun. 2013; in Japanese).
45. ANRE, *Status of introduction of renewable electricity facilities*, http://www.enecho.meti.go.jp/saiene/kaitori/dl/setsubi/201305setsubi.pdf (accessed 2 Sep. 2013; in Japanese).
46. Kitamura, *supra* n. 25, at 132.
47. ANRE, *supra* n. 45, at 1.
48. Facilities for which certification is withdrawn under the RPS Scheme can apply for certification under the FIT Scheme. (The application for withdrawal was due by 1 Nov. 2012.) http://www.hkd.meti.go.jp/hokne/saiene_nintei/seido.pdf (Accessed 5 May 2013, in Japanese).

out of 2,586 MW), although only half of the output of biomass power facilities (1,124 MW out of 2,353 MW).[49]

Other outcomes of the FIT Scheme include activation of renewable energy markets due to entry of various industrial sectors irrelevant to energy business, a positive stance of banks (from megabanks to credit unions) to finance renewable energy businesses, and an expectation of energy-related technological innovation, especially the development of batteries.[50] Various industries such as home appliances, information technology, distribution, construction, and agriculture have been entered. Roofs of stores such as convenience stores and men's apparel chain stores have been used to install PV facilities. In addition, the Japan Agricultural Cooperatives and trading companies work together to install PV facilities on the roofs of livestock barns. This movement has spread throughout the country, including underpopulated areas. Idle land such as industrial parks, golf courses, and agricultural land, roofs of factories, and the tops of landfill sites of industrial waste are utilized. From the investment viewpoint, project finances for a period of 20 years and community-based investment involving local financial agencies and civil funds have been created.

*[2]*    *Current Situation of Renewable Energy Facilities*

The present situation of renewable electricity facilities is described by energy source.[51] As mentioned above, the Japanese government has put a great deal of effort into the promotion of PV power. To date in Japan, most PV power is generated by residential PV systems. Residential PV generation (by facilities less than 10 kW) is rapidly increasing and facilities with total capacity of 1,269 MW had begun operation by the end of FY 2012. Non-residential PV generation (by facilities of 10 kW or more) is also increasing (total capacity of 706 MW by the end of FY 2012); in particular, the construction of solar power mega-plants, which were few in number before the FIT Scheme was enforced, has dramatically expanded with the promoting of plans and construction throughout the country. 107 (189 MW) out of 2,709 (12,487 MW) certified facilities had started generation by the end of FY 2012. In addition, new business such as 'roof lending,' invitations to renewable electricity producers from local government, and public PV facilities are being promoted. These implementations also increase the production of PV-related domestic companies.[52]

Large-scale wind power facilities take four to seven years to start operation since procedures such as a preliminary study and environmental assessment need to be carried out before facilities can be constructed. Thus, the facilities that started operation in FY 2012 are those planned before the FIT Scheme was enforced. Ten to

---

49. ANRE, *Terms pointed out in the previous* committee, http://www.meti.go.jp/committee/chotatsu_kakaku/pdf/011_02_00.pdf (Accessed 5 May 2013, in Japanese).
50. Kitamura, *supra* n. 25, at 132.
51. ANRE *Status of renewable energy after the FIT Scheme is enforced.* (Received material from METI; in Japanese).
52. Japan Photovoltaic Energy Association, *Statistics of shipping volume of PV cells*, http://www.jpea.gr.jp/04doc01.html; http://www.jpea.gr.jp/pdf/t130226.pdf (accessed 5 Jun. 2013; in Japanese).

seventy wind power plants are under construction or the subject of environmental assessment at this moment (as of March 2013). One example of a large-scale wind power plant is that in the Nunobiki highlands of Fukushima Prefecture and the second largest in Japan (65.98 MW with 33 wind turbines). This is an example of converting class-1 agricultural land to a wind farm. This attempt not only realizes compatibility between agriculture and wind power plants but also attracts visitors and increases the sales of local products. Not only on-shore but also off-shore wind power plants have recently begun operation. In Japan, four demonstration projects have now been implemented. Two are fixed wind farms off the coasts of Kitakyushu and Choshi, and the other two are floating wind farms off the coast of Fukushima and Goto Islands in Nagasaki Prefecture. In the case of small-scale wind power, on the other hand, there are few certified facilities, and none have yet begun operation.

Japan has a large amount of geothermal resources (the world's third largest, following the United States and Indonesia), and there is thus a large potential to introduce geothermal power plants. However, because of regulations covering resource-rich areas such as national parks, there have been no newly developed geothermal power plants since 1999. After the relaxation of regulations on geothermal power plants in 2012, new plans for the development of geothermal power plants have been proposed. Consequently, 10 or more projects are being developed. On the other hand, some small-scale geothermal power plants will begin operation in the near future, and hot-spring power generation projects are also planned.

Since it takes three to five years for small- and medium-scale hydropower facilities to begin operation, many hydropower projects are now in the investigation phase for commercialization. In addition, the enforcement of the FIT Scheme encourages the development of hydropower plants by establishing local councils for the development of small- and medium-scale hydropower facilities and the reexamining of projects that were not commercialized previously owing to profitability issues. In addition, there has been an increase in the number of repairs to deteriorated power plants after the enforcement. As a result, it is expected that some small- and medium hydropower plants will begin operation in a few years.

Finally, there are biomass power plants in the planning phase and plants that have been certified although only a few of them have begun operation (as of March 2013).

### [D] Challenges Facing the FIT Scheme and Improvements to the Scheme

Because Japan learned from the system implemented by Germany[53] before establishing the FIT Scheme and because only one year has passed since the Scheme began, no major problems have been reported so far. In Japan, the share of PV power is much larger than that of any other form of renewable electricity, i.e., the shares of other forms of renewable electricity are still low. After the introduction of the FIT Scheme, as

---

53. In the FIT Scheme, measures such as the implementation of a unified surcharge, relaxation of the burden of the surcharge for heavy consumers of electricity, and revision of purchase prices were introduced from the beginning.

described in Table 1.2, the introduction of PV facilities, especially PV facilities other than household facilities, has increased rapidly. However, the growth of use of electricity derived from other energy sources including wind power and geothermal power, which have high growth potential in Japan, is still limited.[54,55] To strongly promote the generation of renewable electricity (especially other than PV-derived electricity), it is necessary to overcome possible challenges and improve the FIT Scheme.

There are four aspects to possible challenges facing the FIT Scheme: the purchase price and long-term goal; grid connections; procedures; and institutions and ministries associated with renewable energy. The following discusses ways to overcome these issues.

*[1]    Purchase Price and Long-Term Goal*

The first challenge is the lack of predictability of the purchase price in the next fiscal year, and the lack of a long-term goal in introducing renewable energy.[56] The lack of predictability in the price acts as a disincentive for business operators to invest in renewable electricity in the future. It is important to consider a predictable purchase price system, set the purchase price according to the latest cost data and knowledge, and publish the cost data frequently to allow business operators to predict the price.[57,58] Additionally, setting an ambitious long-term goal to increase the introduction of renewable energy is important in providing a clear long-term outlook for renewable energy markets, which would encourage investment. A clear outlook for energy markets would strongly assist the investment decision-making of manufacturers of various instruments related to power plants.[59]

*[2]    Grid Connection*

Currently, one of the major barriers preventing producers of renewable electricity from PV from implementing projects is related to grid connection.[60] Unlike the situation in European countries such as Germany and the United Kingdom, electricity companies in

---

54. Yusaku Yano, *Potential of geothermal power generation. Research seminar on geothermal power*, http://www.meti.go.jp/committee/materials2/downloadfiles/g81201a05j.pdf (accessed 2 Jul. 2013; in Japanese).
55. EX Research Institute, Asia Air Survey, Pacific Consultants, and Itochu Techno-solutions *FY 2010 survey on renewable energy potential*, http://www.env.go.jp/earth/report/h23-03/ (Accessed 3 Jul. 2013; in Japanese).
56. Japan Renewable Energy Foundation (JREF), *Achievement and challenges of the first year of the FIT Scheme*, http://www.jref.or.jp/images/pdf/20130628/FIT_press%20release_FIT1year_REV.pdf (accessed 4 Jul. 2013; in Japanese).
57. Yushi Inoue and Yumiko Watanabe, *FIT Schemes for renewable electricity in foreign counties*. Energy and Resources 34 (3), 134–138 (in Japanese) (2013).
58. Institute for Sustainable Energy Policies (ISEP), Renewables Japan Status Report 2013. Nanatsumori Shokan: Tokyo (in Japanese) (2013).
59. JREF, *supra* n. 56, at 3.
60. JREF, *Necessity of priority connection*, http://jref.or.jp/images/pdf/20130225/20130224_final result.pdf (accessed 4 Jul. 2013; in Japanese).

Japan have no obligation to provide priority access to renewable energy. As mentioned above, they can refuse the purchase and connection of renewable electricity because of lack of capacity of transmission lines or the reserve margin being too small.

For example, since the introduction of PV power has been concentrated in Hokkaido, the region is approaching the limit of its connection capacity of renewable energy because of its small-scale electricity grid. There is a risk that the Hokkaido Electric Power Company will increasingly refuse connections. Furthermore, although the potential of renewable energy is high in the Hokkaido area, Hokkaido needs to manage the generated electricity in the area since the capacity of electric power interchange is small (600 MW) between Hokkaido and Honshu (the main island of Japan). However, its capacity to control power is small owing to the small demand in the area.[61] In April 2013, the METI released methods for responding to the problems faced by Hokkaido, and decided to (1) revise conditions of connections in the specific area to expand the capacity of connection, (2) introduce large-scale batteries to expand the connection of renewable electricity and (3) expand wide-area grid operation according to electricity system reform.[62]

Grid connection is a major problem not only for PV power but also for other forms of renewable electricity such as wind power. In Japan, areas where the wind speed is sufficient for constructing wind farms are limited to parts of Hokkaido and Tohoku.[63] Since the population density is low in these areas, the infrastructure of power grids is limited in these areas. This could result in some of the electricity produced by wind power generation being unable to be fed into the grid. Therefore, the areas suitable for wind power generation in Hokkaido and Tohoku are designated as strategic areas, and it is planned to support private business in developing grids and to demonstrate the business model and technological issues. In particular, a special-purpose company has been established (over half the investment for which is from private businesses and the remaining is from electricity companies) and the investment is recouped through use fees paid by wind power producers. Since this business is unprofitable, half of the project cost is subsidized by the government.

To enhance the generation of renewable electricity, it is important to address grid connection problems, such as by obliging electricity companies to provide priority access,[64,65] and by expanding the infrastructure of energy transmission and distribution.[66] In addition, large-scale batteries for the major power grids of renewable

---

61. METI, *Summary of the press conference by the Minister of METI, Yukio Edano*, http://www.meti.go.jp/speeches/data_ed/ed121207j.html (accessed 9 May 2013).
62. ANRE, *Response to the connection of large-scale PV in Hokkaido*, http://www.meti.go.jp/press/2013/04/20130417003/20130417003.pdf (accessed 6 Jul. 2013; in Japanese).
63. ANRE, *Basic information for the Procurement Price Calculation Committee*, http://www.meti.go.jp/committee/chotatsu_kakaku/pdf/008_02_00.pdf (accessed 9 May 2013; in Japanese).
64. Institute for Sustainable Energy Policies (ISEP), Renewables Japan Status Report 2013. Nanatsumori Shokan: Tokyo (in Japanese) (2013).
65. Japan Renewable Energy Foundation (JREF), *Achievement and challenges of the first year of the FIT Scheme*, http://www.jref.or.jp/images/pdf/20130628/FIT_press%20release_FIT1year_REV.pdf (accessed 4 Jul. 2013; in Japanese).
66. Minoru Nakamura (Toyo Keizai), *A lot of challenges of renewable energy*, http://toyokeizai.net/articles/-/13636 (accessed 5 Jul. 2013; in Japanese).

electricity facilities themselves will be needed in preparing for a rapid increase in the generation of renewable electricity. It takes a decade to develop power transmission lines; however, the capacity for the reception of renewable electricity is limited. In FY 2012, JPY 29.59 billion was allocated to demonstration projects of large-scale batteries. In FY 2013, the government's budget contained JPY 2.7 billion of subsidies for battery R&D[67] to reduce the unit cost of a battery (JPY 40,000/kW) to the level of pumped hydropower generation (JPY 23,000/kW).

### [3] Procedures

Unlike the case for PV power generation, one of the most significant problems facing wind and geothermal power generation is that it requires environmental assessment, which takes around three to four years.[68] Thus, acceleration and simplification of the process need to be considered, and the keys are (1) to shorten reviews by national and local governments, and (2) to shorten environmental investigation of the FIT Scheme. Aiming to halve the proceeding period of environmental assessment for wind and geothermal power plants, the METI and Ministry of the Environment (MOE) decided to implement parallel reviews by national and local governments and to rationalize the reviews through the participation of business operators. In addition, it is considered necessary to collect and manage information used in previous environmental assessments for the benefit of business operators in their assessment.

### [4] Institutions and Ministries regarding Renewable Energy

To promote the generation of renewable electricity, it is necessary to link and fill the gaps among institutions and ministries in overcoming renewable energy-related problems. Table 1.3 gives the related Acts and ministries, and the challenges facing the promotion of renewable electricity. The table also gives the options for overcoming the challenges and the current actions being carried out to do so.[69]

---

67. Kitamura, *supra* n. 25, at 133.
68. ANRE, *Basic information for the Procurement Price Calculation Committee*, http://www.meti.go.jp/committee/chotatsu_kakaku/pdf/008_02_00.pdf (accessed 9 May 2013; in Japanese).
69. ANRE *Status of renewable energy after the FIT Scheme is enforced*. (Received material from METI; in Japanese).

*Table 1.3  Institutional Challenges Facing the Promotion of Renewable Electricity*

| Type | Act | Main Implementing Ministries | Challenges in Promoting Renewable Electricity | Options to Overcome the Challenges | Current Actions to Overcome Challenges |
|---|---|---|---|---|---|
| Location regulations | Factory Location Act | METI | Greening regulation in the act is applied even if PV facilities are installed on the roof of non-factory buildings such as business establishments. | Revision of regulations on area and greening of manufacturing facilities related to PV facilities. | PV facilities are exempted from the notification in the act. In addition, they are classified as environmental facilities. |
| | Agricultural Land Act /Act on Establishment of Agricultural Promotion Regions | MAFF | Conversion of large-scale agricultural land (e.g., class-1 agricultural land) is not permitted by renewable electricity producers except the nine electricity companies. | Revision of regulations related to agricultural land. | Treatment of installation of renewable electricity facilities when they do not have adverse effects on secure agricultural land and contribute to regional development has been clarified. |
| | Act on Utilization of National Forest Land/ Public Accounting Act | FA MOF | Lending of national forest by private contract to renewable electricity producers is not permitted except for the nine electricity companies, because their businesses are not for the public. | Lifting the ban on lending national forest land to private power producers and companies of steam production for geothermal power. | Certified facilities in the FIT Scheme are added to the coverage of private contracts. |

| Type | Act | Main Implementing Ministries | Challenges in Promoting Renewable Electricity | Options to Overcome the Challenges | Current Actions to Overcome Challenges |
|---|---|---|---|---|---|
| | Forest Act | FA | Procedures of unspecifying protected forest can be rejected by local governments or national government that strictly require evidence that no other suitable places for development exist. | Unspecifying protection forests and concretization of rules on work permits in protected forests. | Considerations relating to operation of the protected forest regulation are released. |
| | Natural Parks Act | MOE | Establishment of geothermal power plants is virtually prohibited in special areas. | Lifting the ban on mining investigation and development of geothermal power in the special areas of national parks. | Notification to permit the development of class-2 and 3 special areas with conditions has been released. |
| | Hot Springs Act | MOE | It is difficult to obtain mining permits, because the criteria for mining permits are unclear. | Mining permits based on scientific evidence. | A guideline allowing scientific reviews has been released. |

| Type | Act | Main Implementing Ministries | Challenges in Promoting Renewable Electricity | Options to Overcome the Challenges | Current Actions to Overcome Challenges |
|---|---|---|---|---|---|
| | River Act | MLIT | Procedures for small-scale hydropower are the same as those for large-scale hydropower. | Simplification of procedures for permitting water rights related to small-scale hydropower. | Procedures for small-scale hydropower have been rationalized and simplified (according to used water flow and the scale of power generation). A license system has been introduced for subordinate power generation in the range of existing water right permits. |
| Safety regulations | Building Standards Act/ Ship Safety Act | MLIT | Building structural standards required for off-shore wind power are unclear. | Improvement of legal systems related to off-shore wind power (e.g., structural standards). | Technological standards have been established according to the Ship Safety Act. Structural standards for floating wind power facilities have been unified with the Ship Safety Act (and exempted from the Building Standards Act). |

| Type | Act | Main Implementing Ministries | Challenges in Promoting Renewable Electricity | Options to Overcome the Challenges | Current Actions to Overcome Challenges |
|---|---|---|---|---|---|
| | Electricity Business Act | METI | The safety regulations are excessive for PV facilities, because they are safer than other power generation facilities. | Relaxation of safety regulations related to PV. | The ministerial ordinance has been revised to expand the range in which the notification of construction plans is unnecessary. The capacity range for which it is not necessary to appoint a chief electricity engineer has been expanded (from less than 1,000 kW to less than 2,000 kW). |
| Grid and other regulations | Electricity Business Act | METI | Predictability of possibility and costs of grid connection is not high. | Smoothing grid connection. | Revisions have been implemented, such as disclosure of possible points for grid connection and details of connection costs and work periods, unionization of documents for the procedure, and shortening of the standard processing periods. |

| Type | Act | Main Implementing Ministries | Challenges in Promoting Renewable Electricity | Options to Overcome the Challenges | Current Actions to Overcome Challenges |
|---|---|---|---|---|---|
| | Waste Management and Public Cleansing Act | MOE | If biomass fuel is treated as waste, it will be costly because permissions for waste power facilities/ business and transportation is required. | Clarification of treatment of biomass fuel in the act. | Cases in which local governments do not consider biomass fuel as waste have been collected. It has been clarified that power producers are exempted from regulations of the Waste Management and Public Cleansing Act in the case that the transportation costs of biomass fuel are higher than the sale value. |

\* MAFF: Ministry of Agriculture, Forestry and Fisheries, FA: Forestry Agency, MOF: Ministry of Finance, MLIT: Ministry of Land, Infrastructure, Transport and Tourism.[70]

So far, institutional arrangements for renewable energy have been mainly designed and discussed by the METI. However, in promoting renewable energy, Acts implemented by other ministries, as described in Table 1.3, will also be related. For example, in introducing geothermal power, there are conflicts with the Act on Utilization of National Forest Land and Public Accounting Act implemented by the Forestry Agency and Ministry of Finance, and the Natural Parks Act implemented by the MOE. Therefore, it is important to identify conflicts and lessen the gaps among the Acts and ministries.

---

70. ANRE *Status of renewable energy after the FIT Scheme is enforced*. (Received material from METI; in Japanese).

## §1.04 CONCLUDING REMARKS

This chapter presents an overview of the effects and challenges of previous and current renewable energy-related policies and institutions in Japan, with a focus on the newly-introduced FIT Scheme.

The RPS Scheme and Net-metering Scheme for Photovoltaic Power have different scopes and methods, and the latter was more effective in terms of increasing PV power generation. Since the FIT Scheme was launched in 2012, the Japanese government has been trying to unify renewable energy-related institutions under it, including integrating the RPS Scheme and the Net-metering Scheme for Photovoltaic Power. Additionally, subsidies for renewable energy are being reduced. The FIT Scheme has a large potential to increase the use of renewable electricity and renewable energy more generally by allowing renewable energy business operators to develop long-term investment plans and obtain financing from banks. However, as described in the previous section, so far, the FIT Scheme has only really affected PV power introduction. There are possible obstacles preventing the FIT Scheme from enhancing the use of renewable energy dramatically, including the lack of a predictable purchase price and long-term goal for renewable electricity, limited grid connections and connection refusals, ineffective procedures, and gaps among institutions and ministries associated with renewable energy.

Since the Fukushima Nuclear Accident and the resultant shutting down of nuclear power plants across Japan, there has been a growing demand to increase the use of renewable electricity in Japan. Additionally, the Bill of the Basic Act on Global Warming Countermeasures included a target of raising the share of renewable energy to 10% of the total primary energy supply by 2020, and the Energy and Environmental Council of the Japanese government decided in September 2012 to aim for zero nuclear power generation by the 2030s. Although these two measures were scrapped after the dissolution of the lower house, the Democratic Party of Japan has contributed to raising public awareness of the need for renewable electricity to some degree.

To enhance the generation of renewable electricity in Japan, it is important to overcome the challenges of the FIT Scheme including the lack of predictability of the purchase price in the next fiscal year and the lack of a long-term goal in introducing renewable energy, grid connections-related issues, long lead-in period for wind and geothermal power generation, and gaps among institutions and ministries associated with renewable energy. In addition, it is necessary to raise public awareness of the need for renewable electricity. Although the Fukushima Nuclear Accident and the shutting down of nuclear power plants across Japan has raised public awareness, whether we can keep doing so is highly dependent on the policy direction of the current regime of the Liberal Democratic Party Shinzō Abe.

CHAPTER 2
# From FIT to RPS under the Low-Carbon Green Growth Initiative: Moving Forward or Backward for the Expansion of Renewable Energy in Korea?*

*Deok-Young Park & Taehwa Lee*

## §2.01 INTRODUCTION

For many years, the energy policy of the Republic of Korea mainly focused on providing a stable and reliable energy supply with low prices in order to maintain the global competitiveness of Korean companies and promote economic growth. However, this old paradigm has been replaced by a new energy paradigm, which reflects current conditions. Higher oil prices; increased environmental awareness at local, regional and global levels; concerns on energy supply security; and a rise of environmental Non-Governmental Organizations (NGOs) have jointly created movement from an energy supply oriented paradigm to a new energy paradigm that focuses on demand

---

\* 'This manuscript was published for the Korean Journal *Modern Society and Public Administration* (Vol 23., No. 3) in December 31, 2013. This work was supported by the National Research Foundation of Korea Grant funded by the Korean Government (N R F - 2 0 1 3 S 1 A 3 A 2 0 5 4 9 6 9 ).'

side management and the supply of environmentally sustainable energy in Korea.[1] To respond to these issues, the Korean government has developed and implemented many new environmentally-friendly laws in the energy sector since 2006.[2] Furthermore, 'Low Carbon Green Growth,' a new national vision proclaimed in 2008, became the higher threshold which strongly increases the use of new and renewable energy (hereafter, NRE).

While announcing the green growth initiative, the Korean government has also established an 11% target of NRE in Total Primary Energy Supply(TPES) by 2030 in the Third Renewable Basic Plan. Korea replaced its Feed-in Tariff (FIT) program with a renewable portfolio standard (RPS) program in 2012 in order to meet its NRE targets.[3] The Korean government expressed its ambitious goal to expand NRE source in its Sixth Electricity Supply Demand Plan, which was announced in February 2013. According to the Plan, the share of renewable energy sources in Korea's total electricity supply is likely to rise to 12% by 2027.[4]

Despite Korea's efforts, the contribution of renewable sources to TPES in Korea is still the lowest among Organisation for Economic Co-operation and Development (OECD) member countries.[5] The FIT program implemented between 2002 and 2011 and the first year of the implementation of the RPS program were attempts to utilize policy tools for NRE expansion. Thus, the increase in NRE supply in recent years and the government's ambitious goal to expand NRE source in the future requires careful examination. This chapter is designed for this purpose. First, the chapter briefly reviews literature on the FIT and RPS programs. Second, it explains the FIT program of Korea and analyzes the RPS program in detail. Third, it analyzes the limits and potential of the current RPS program in Korea. The chapter concludes with legal and institutional insights that need to be further considered for the successful implementation of the RPS program.

---

1. Heoseok Kim, Eui-soon Shin and Woo-jin Chung, *Energy Demand and Supply, Energy policies, and Energy Security in the Republic of Korea*, 39 Energy Policy 6882, 6887 (2011).
2. An important law fostering the use and deployment of NRE in Korea is the Act on the Promotion of the Development, Use and Diffusion of New and Renewable Energy. The act was introduced in 1997 and revised various times including the last revision in March, 2013. NRE in the Act refers to renewable energy, including solar energy, bioenergy, wind energy, water energy, marine energy, geothermal energy, and the energy converted from existing fossil fuels including liquefied or gasified coal and from gasified heavy residual oil, hydrogen energy, and energy prescribed by Presidential Decree. The categorization of this NRE does not include petroleum, coal, nuclear power or natural gas. See Act on the Promotion of the Development, Use and Diffusion of New and Renewable Energy, Art. 2 (23 Mar. 2013). Critics argue that NRE in the current categorization is not environmentally sustainable and not strictly renewable since liquefied or gasified coal is the energy converted from existing fossil fuels. See Sun Jin Yun, *Korea's Energy System and Sustainability: Focused on an Analysis of Continuance of Unsustainability*, 78 Economy and Society 12, 15 (2008).
3. International Energy Agency (IEA), *Energy Policies of IEA Countries: The Republic of Korea 2012 Review* (IEA Publications 2012).
4. Ministry of Knowledge Economy, *The 6th Basic Plan for Long-term Electricity Supply and Demand (2013–2027)*. (MKE Notice #2013-63). (Ministry of Knowledge Economy, 2013).
5. IEA, *supra* n. 3, at 93.

## §2.02 LITERATURE REVIEW ON FIT AND RPS PROGRAMS

FIT and RPS are two major policy tools that encourage the use and deployment of NRE sources for electricity generation. It is considered that the RPS program is more compatible with the market regime, while the FIT program is better suited for controlling environmental effectiveness, collective cost and consumers' cost.[6] The FIT is also known as one of the most effective policy instruments since it reduces the burden of the cost barriers for using renewable energy. Under FIT, the utility pays a rate or tariff that is fixed for a 15–25 year period. The rate is decided by the government to ensure that the renewable energy producer can gain a reasonable economic return on his or her investment and therefore continue to do his or her research and invest in renewable energy in order to reduce the cost of energy over time.[7] The guaranteed access to the grid, favorable rate per unit and the tariff term guarantee encourage producers, manufacturers, investors and suppliers to participate in the installation of renewable energy systems since there is no business risk involved.[8]

Despite these benefits of using FIT as an effective policy tool, there are certain obstacles in implementing FITs. First, decisions on payment amounts for individual technologies, payment structures (such as fixed or declining), and payment duration involve guesswork by policymakers on future market conditions and rates of technological improvements. Once specific price paths, including level, structure, and duration, are guessed significantly and decided upon by the policymakers, changing those paths becomes difficult and costly.[9] Second, the FIT program generally places a financial burden on the government through the process of supporting the financial costs of the renewable energy generators.[10] Third, FIT with time-constant tariffs cannot sufficiently control collective cost and prevent generators' rent-seeking behaviors.[11]

Under RPS, the government mandates that a certain percentage of electricity production must be generated from renewable energy sources. Under the general mechanism of RPS, renewable energy generators obtain certificates, which is mandatory, for the electricity they generate, and can sell these to supply companies. Supply

---

6. Dominique Finon, *The Social Efficiency of Instruments for the Promotion of Renewable Energies in the Liberalised Power Industry*, 77 Annals of Public and Cooperative Economics 309–343 (2006).
7. David Grinlinton and LeRoy Paddock, *The Role of Feed-In Tariffs in Supporting the Expansion of Solar Energy Production*, 41 University of Toledo Law Review 943, 944 (2010).
8. Jong Yeong Yi, Kee Bong Yoon and Won Seog Park, *The Renewable Portfolio Scheme*, 35 Environment Law Review 279, 286 (2013). *See also* Grinlinton and Paddock, *supra* n. 6, at 944.
9. Jonathan A. Lesser, Xuejuan Su, *Design of an Economically Efficient Feed-In Tariff Structure for Renewable Energy Development*, 36 Energy Policy 981, 982 (2008).
10. Tae Eun Kim, *The Study on the Response to the Dilemma as Institutional Change and the Alternative Factor: Focusing on the FIT Scheme*, 43 Korean Public Administration Review 179, 200 (2009).
11. Finon, *supra* n. 5, at 337.

companies then pass on the certificates to the regulatory body to demonstrate their compliance with their regulatory obligations. Since it is based on a market mechanism, the RPS relies on the private market for its implementation. The RPS program is known to have the following benefits: First, it provides market incentives for generators to reduce development costs. Second, the financial burden is shifted from the government to retail electricity suppliers.[12] But the RPS program, like the FIT program, also has its own shortcomings. First, the possibility of rent-seeking behaviors by electricity generators from renewable energy sources still exists.[13] Second, due to the investment uncertainty in the RPS program, newcomers and small scale generators are discouraged from entering the market. The RPS program only applies to electricity-based technologies, while non-electric technologies are not supported.[14] Furthermore, even though the RPS program tries to support diverse technologies, energy suppliers who participate in the RPS program usually favor the least-cost technology, which ultimately decreases diversity in renewable energy technologies.[15]

## §2.03 THE FIT PROGRAM IN KOREA

Korea's TPES in 2011 was estimated to be 258 million tons of oil equivalent (Mtoe). TPES had increased 3% compared to 2010, and 37% compared to 2000. Economic growth during those 10 years was the main driver for this increase. Oil (36%), coal (31%), natural gas (16.2%), and nuclear energy (15.2%) are the four main contributors to the energy supply, while the share of renewable energy to TPES is almost negligible. Total final consumption (TFC) of energy in 2010 was 157.4 Mtoe, an increase of 6.5% compared to 2009 and of 23.8% compared to 2000. Real GDP grew by over 4% per year on average during the same period. The industry sector took the largest share of TFC, representing around 52.3%, while the transportation sector and residential sector took 19.4% and 12.6% respectively in 2010.[16]

TPES in Korea increased from 93 Mtoe to 258 Mtoe between 1990 and 2011. This was more than a 170% increase during that period. At the same time, renewable energy increased from only 1.3 Mtoe to 4.0 Mtoe, which was more than a 200% increase over

---

12. Kim, *supra* n. 9, at 200.
13. Tae Hyung Kwon, *Regulating Rents from Renewable Energy Policies: Focusing on RPS in South Korea*, 11 Korean Energy Economic Review 141, 147 (2012).
14. Catherine Mitchell, Peter Connor, *Renewable Energy Policy in the UK 1990–2003*, 32 Energy Policy 1935, 1940–1941 (2004).
15. Melanie Grant, *Where Are They Now? A Look at the Effectiveness of RPS Policies*, 2011 Brigham Young University Law Review 849, 852 (2011).
16. IEA, *supra* n. 3, at 17–18.

the same period. The largest share of NRE supply came from biofuels and renewable waste. They represented almost 66.7% of NRE production in 2011, while hydropower represented 21.6% and the remainder consisted of solar photovoltaic (PV) and wind. NRE contributed to 7,849 GWh in electricity supply in 2011. Sixty percent came from hydro power. Even though the share of wind and solar PV in electricity supply is still very low, recent years show significant growth in both sectors.[17] The share of NRE in primary energy supply increased from 1.2% to 2.4% between 2001 and 2011.[18]

This recent growth resulted from the implementation of the FIT program after the revision of the Act on the Promotion of the Development, Use and Diffusion of Alternative Energy in 2002 to initiate the FIT program. The government compensated electricity generators for the differences between the costs of electricity generated from NRE sources and fossil-fuel based energy sources to promote the production, use, and deployment of renewable energy sources. Such compensation is referred to as feed-in-tariff (FIT). The period and volume are the main characteristics of the FIT program. The FIT program guaranteed 15 years of support for all new and renewable electricity facilities, while the total number of years for the support of PV can be either 15 or 20 years.[19] The standard prices for NRE have been modified several times after being formulated in 2002. The FIT program provided a window of opportunity for active investment on NRE. Certain NRE sources such as PV and wind had received special benefits for their increased market shares. The annual total volume of PV dissemination was just over 200 kW before 2004. However, with the beginning of the FIT program and following recognition from investors, PV installation increased dramatically and became 497 MW in 2011.[20] Table 2.1shows the amount invested by the Korean government for the FIT program from 2002 until the end of August in 2011 and the amount of electricity generated by the FIT program from NRE.[21]

---

17. Ibid., 93.
18. Korea Energy Economics Institute, *Yearbook of Energy Statistics*, 5 (Korea Energy Economics Institute, 2012).
19. IEA, *supra* n. 3, at 97.
20. Korea Energy Management Corporation (KEMCO), *Program for promoting NRE utilization*, http://www.kemco.or.kr/new_eng/pg02/pg02040700.asp (accessed 14 Jul. 2013).
21. Nam Hun Kang, 'The New and Renewable Energy Law: Focusing on RPs', in Hong Sik Cho and others (eds), *Energy Law and Policy in the Age of Climate Change* (Parkyoungsa, 2013) 339.

Table 2.1 General Information on NRE Generation under FIT in August 2011[22]

| Type | Hydro | Wind | PV | Fuel Cells | LFG | Biogas | Biomass | Wastes | Total |
|---|---|---|---|---|---|---|---|---|---|
| Generation Capacity (kW) | 87,396 | 320,205 | 496,624 | 50,500 | 74,868 | 2,711 | 5,500 | 2,247 | 1,040,096 |
| Number of Facilities | 63 | 15 | 1,991 | 20 | 14 | 3 | 1 | 1 | 2,108 |
| Generation (MWh) | 2,212,603 | 3,575,960 | 2,322,794 | 746,091 | 2,625,610 | 36,070 | 51,263 | 12,862 | 11,583,253 |
| Financial Support (Million KRW) | 27,667 | 26,152 | 1,168,098 | 103,565 | 16,517 | 344 | 230 | 61 | 1,342,634 |

22. Yi, Yoon and Park, *supra* n. 7, at 288.

Despite its great contribution of disseminating NRE, the FIT program created some problems, which finally led to the shift to the RPS program in Korea. First, the FIT program failed to fairly distribute investment for diverse renewable energy sources. For example, PV received 88.2% of the entire FIT financial support in 2010.[23] Second, the FIT program experienced difficulty in meeting the targets for NRE since the government only had a limited financial support for the FIT program. Third, financial support for the compensation for the differences between the costs of electricity generated from NRE sources and fossil-fuel based energy sources came from the Electric Industry Foundation Fund, most of which was spent for the FIT program. Because the Fund is collected from the electricity bills that consumers pay, the FIT program is inconsistent with the 'polluter pays' principle in that consumers who use environmentally friendly energy ultimately pay for the costs through their electricity bills. Fourth, because of the relationship between the Fund and the FIT program, as more NRE generators participated in the program, the government's financial burden increased, which eventually led it to stop the program itself.[24] Additionally, the excessive political interference, the lack of an institution for mediation and consultation for policy implementation, and absence of accountability and transparency also contributed to the early shift from the FIT program to the RPS program.[25]

## §2.04 THE RPS PROGRAM IN KOREA SINCE 2012

### [A] The General Structure of the RPS Program

Section 5 of Article 12 of the Act on the Promotion of the Development, Use and Diffusion of New and Renewable Energy defines the RPS program as a system requiring 'mandatory generators' to 'mandatorily supply not less than a certain percentage of the volume of electricity generation by using new and renewable energy.'[26] The Act also mandates that the total volume of power generation by a generator shall be prescribed by Presidential Decree on a yearly basis within 10% of the total volume of electricity generation.[27]

An Enforcement Decree of the same Act defines mandatory generators to be power generators with a power generation capacity of 500 MW or more (excluding NRE generation capacity), K-Water and Korea District Heating Corporation.[28] These cover all 13 publicly-owned and privately-owned power generators including K-Water, Korea District Heating Corporation, Korea Hydro & Nuclear Power Company (KHNP), Korea

---

23. Ibid., 288.
24. Ibid., 287–288.
25. Min Gyo Koo, *South Korea's Feed-in Tariff Program and Its Implications for New Industrial Policy: Policy Design, Implementation, and Learning*, 22 Korean Public Administration Review 1, 20 (2013).
26. Act on the Promotion of the Development, Use and Diffusion of New and Renewable Energy, Art. 12(5) (23 Mar. 2013).
27. Ibid.
28. Enforcement Decree of the Act on the Promotion of the Development, Use and Diffusion of New and Renewable Energy, Art. 18(3) (23 Mar. 2013).

East-West Power (KOEWP), Korea South-East Power (KOSEP), Korea Midland Power (KOIMPO), Korea Western Power (KOWEPO), Korea Southern Power (KOSPO), GS E&S, GS Power, POSCO Power, MPC Yoolchon, and MPC Daesan.[29] The renewable energy portfolio to be provided by mandatory power generators starts from 2% from renewable energy sources in total power generation in 2012. It increases by 0.5% every year until 2016, then it increases by 1% annually from 2016 until 2022 when the renewable energy portfolio reaches 10%. The RPS ratio is subject to adjustment every three years based on a review by the Minister of Industry, Trade and Energy[30] on technological development, supply targets for NRE, performance level, and other circumstances.[31]

The portion of renewable energy to be generated by mandatory generators starts from 2% in 2012 and increases by 0.5% each year until 2016. It will be increased by 1% from 2017 until 2022, on which the portfolio will reach 10%. The Korean RPS program has solar carve-out provisions that require mandatory power generators to generate certain amounts of their power from solar energy.[32] Section 5(2) of Article 12 of the Act states, 'with respect to new and renewable energy which requires balanced use and distribution, it is allowed to supply part of the total volume of mandatory supply using such new and renewable energy, as prescribed by Presidential Decree.'[33] In order to support and deploy solar energy, the Korean government chose solar energy as a new and renewable energy, which needs the 'balanced use and distribution' as indicated in the Act. The amount of power generated from solar power is planned to rapidly increase for five years, beginning from 2012 until 2016.[34] The obligatory amount of power generated from solar energy in the RPS program begins with 1,577 Gwh in 2012, 723 Gwh in 2013, 1,156 Gwh in 2014, and 1,577 Gwh after 2015.[35]

A mandatory generator that has a facility with the capacity of 5 GW or more has to purchase not less than 50% from a generator whose facility does not have the capacity of 5 GW or more.[36] The purpose of this rule is to promote small-scale PV generators because of concerns about the possibility of the RPS seriously undermining any incentive to invest in photovoltaic power.[37]

---

29. KEMCO New and Renewable Energy Center, *RPS Program*, http://www.knrec.or.kr/knrec/12/KNREC120700_02.asp (accessed 11 Jul. 2013).
30. After passing the revision of Government Organization Act in the National Assembly on 22 Mar. 2013, The Ministry of Knowledge Economy was restructured as the Ministry of Industry, Trade and Energy (MOTIE).
31. Enforcement Decree of the Act on the Promotion of the Development, Use and Diffusion of New and Renewable Energy, Art. 18(4) (23 Mar. 2013).
32. The scope of solar energy is limited to the method that generates electricity from converting sunlight. KEMCO New and Renewable Energy Center, *supra* n. 28.
33. Act on the Promotion of the Development, Use and Diffusion of New and Renewable Energy, Art. 12(5)-2 (23 Mar. 2013).
34. Enforcement Decree of the Act on the Promotion of the Development, Use and Diffusion of New and Renewable Energy, Art. 18(4) (23 Mar. 2013).
35. *Ibid.*
36. The Ministry of Knowledge Economy, *RPS Management and Operation Directive*, Art. 10 (25 Jun. 2012). (MKE Notice #2012-134).
37. Do-Yo Kim, *Introduction of RPS and Phase-out of FIT in Renewable Energy Policy*, http://www.iflr.com/Article/3072471/Introduction-of-RPS-and-phase-out-of-FIT-in-renewable-energy-policy.html (accessed 11 Jul. 2013).

The Korean RPS Program also adopts a flexible mechanism. A mandatory generator may defer up to 20% of the performance of a duty against the amount of renewable energy that they are required to provide to the following year. Generators are allowed to defer up to 30% of their duty on energy generation from renewable sources by 2014.[38] The RPS program in Korea also imposes a penalty surcharge up to an amount calculated by multiplying 150% of the average market price of a Renewable Energy Certificate (REC) of the relevant year if a generator does not supply a required mandatory amount of energy from a NRE source.[39]

### [B]   Issuance and Trading of REC

The RPS program in Korea allows the government-affiliated institutions to issue and trade RECs. A generator who provides energy from a NRE source may obtain a certificate certifying the fact that proves the provision of NRE from a certified institution by the Minister of Industry, Trade and Energy.[40] The Korean New and Renewable Energy Center (KNRE) of the Korean Energy Management Corporation under the Ministry of Trade, Industry, and Energy (MOTIE) is responsible for the issuance and trading of the certificates.[41]

KNRE should issue an REC after checking the supplying amount and the supplying period of each NRE. For the targeted NRE for the balanced use and deployment and for the purpose of promoting technological development, RECs can be issued for the amount that the real supplying amount is multiplied with a weighted value (multiplier) of each NRE.[42] As shown in Table 2.2, there are various multipliers depending on the type of NRE sources. The multipliers are given according to various factors including impacts on the environment, technological development and industry revitalization, generation cost, energy potentials, and impacts on reducing the emission of greenhouse gases.[43]

---

38. Enforcement Decree of the Act on the Promotion of the Development, Use and Diffusion of New and Renewable Energy, (Presidential Decree # 22382), Supplementary Provisions Art. 3 (23 Mar. 2013).
39. Act on the Promotion of the Development, Use and Diffusion of New and Renewable Energy, Art. 12(6)-1 (23 Mar. 2013).
40. *Ibid.*, Art. 12(7).
41. The Ministry of Knowledge Economy, RPS Management and Operation Directive, Art. 15(3) (25 Jun. 2012). (MKE Notice #2012-134).
42. Act on the Promotion of the Development, Use and Diffusion of New and Renewable Energy, Art. 12(7) (23 Mar. 2013).
43. Enforcement Decree of the Act on the Promotion of the Development, Use and Diffusion of New and Renewable Energy, Art. 18(9).

*Table 2.2   Multipliers of NRE[44]*

| Grouping | Multiplier | Eligible Resources | | |
| --- | --- | --- | --- | --- |
| | | Installation Type | Land Type | Capacity |
| Solar Energy | 0.7 | In case of not using buildings and existing facilities | 5 areas (rice field, dry field, orchard, pasture, forest land) | |
| | 1.0 | | Other 23 areas | Over 30 kW |
| | 1.2 | | | Under 30 kW |
| | 1.5 | Buildings and existing facilities, Installing facilities on the surface of the water | | |
| Other NRE | 0.25 | Integrated Gasification Combined Cycle(IGCC) and By-Product gas | | |
| | 0.5 | Wastes and landfill gas | | |
| | 1.0 | Hydro, onshore wind, biogas, biomass, Refuse-Derived Fuel(RDF), tidal (with tide embankment) | | |
| | 1.5 | Wood biomass, offshore wind (under 5 km) | | |
| | 2.0 | Offshore wind (over 5 km), tidal (without tide embankment), fuel cell | | |

The term of validity of RECs will be three years from the issuance date. However, if a generator submits the RECs for his or her performance of mandatory REC requirements, those RECs are considered to be ineffective.[45]

The RECs may be issued to NRE facilities beginning their commercial operations after 1 January 2012. However, in order to promote solar energy market, NRE generation from solar energy facilities under the Renewable Purchase Agreement (RPA) may be issued RECs.[46] The RECs can be also issued for the following facilities: (1) NRE facilities passing inspections required under the Article 63 of the Electric Business Act after 17 September 2010, (2) Hydro power facilities over 5,000 kW capacity, (3) NRE facilities receiving support from FIT, (4) New and renewable energy facilities not supported by FIT among businesses operated under the RPA, (5) By-product power facilities passing inspections under the Article 63 of the Electric Business Act before 31 December 2011 after obtaining business permits under the

---

44. The Ministry of Knowledge Economy, RPS Management and Operation Directive, asterisk # 3 (25 Jun. 2012). (MKE Notice #2012-134).
45. Act on the Promotion of the Development, Use and Diffusion of New and Renewable Energy, Art. 12(7) (23 Mar. 2013).
46. The RPA is a transitional program with the purpose of expanding and distributing solar energy for RPA contractors from 2009 until 2011. The RPA contractors are KHNP, KOEWP, KOIMPO, KOWEPO, KOSEP, and KOSPO.

Article 7 of the Electric Business Act before 12 April 2000, and (6) NRE facilities being certified as buildings, which use certain amounts of their energy from NRE sources.[47]

A generator may be restricted from receiving an REC in the following cases. NRE facilities receiving financial supports from the State or local governments may be restricted from receiving RECs. For example, a generator receiving a financial support from FIT may not receive RECs.[48] A generator should request an issuance of the REC not later than 90 days from the date when he or she supplies NRE. And the KNRE should issue the RECs to the generator who requests issuances of RECs within 30 days of his or her application.[49] Korea Power Exchange (KPX) is responsible for the creation and operation of the trading market. For trading RECs, the Over-the-Counter (OTC) markets and the spot markets are utilized. OTC markets are markets where 'seller and purchaser enter into a sale and purchase agreement and report the contract to the KNRE for the transfer of the RECs', while the spot markets are markets where 'the sales are executed via auction or tendering.' The spot markets are divided into RECs from solar power and RECs from NRE resources other than solar power.[50]

The current RPS program prohibits REC purchasers from reselling the RECs.[51] Financial profits generated from trading of RECs may be allocated for supporting the Electric Industry Foundation Fund.[52] The trading of RECs may be restricted in the following cases. First, the trading of RECs is restricted if the REC is issued for hydropower facilities with capacity exceeding 5,000 kW. Second, the trading of RECs is restricted in the case of issuing REC from IGCC or wastes that were generated from generation facilities committing unlawful activities.[53]

## §2.05 CHALLENGES AND SOLUTIONS

### [A] Challenges

After shifting from the FIT program to the RPS program, 64.7% of the mandatory generators' requirements were met in the first year of implementing the RPS program. Only 26.9% was fulfilled by the generators' own NRE facilities while the remainder was satisfied by buying RECs.[54] Generators who did not fulfill their requirements fear that they cannot meet their targets and thus have to be heavily financially burdened

---

47. The Ministry of Knowledge Economy, *RPS Management and Operation Directive*, Art. 6(1)-6 (1 Mar. 2013). (MKE Notice #2013-48).
48. Act on the Promotion of the Development, Use and Diffusion of New and Renewable Energy, Art. 12(7) (23 Mar. 2013). Even if the shift from the FIT program to the RPS program in 2012, generators who participated in the FIT program before 2012 still receive the financial support from the government.
49. Enforcement Decree of the Act on the Promotion of the Development, Use and Diffusion of New and Renewable Energy, Art. 18(8).
50. Kim, *supra* note 36.
51. New and New and Renewable Energy Center Notice #2012-5.
52. Enforcement Decree of the Act on the Promotion of the Development, Use and Diffusion of New and Renewable Energy, Art. 18(7).
53. *Ibid.*, Art. 2(2).
54. Yoon Seok Park, *The Mixed Reality of the RPS Program for One Year*, Electric Power 42 (2013).

with penalties. There is also a voice from solar industry that the solar carve-out program itself hinders companies from increasing their investments in PV since companies have tendencies to stay within their targets.[55] The following section explores diverse changes to go beyond the limits of the current RPS program in order to make it successful in expanding renewable energy in Korea.

### [B] Solutions: How to Improve the Effectiveness of RPS

#### *[1] Multiplier and Public Participation*

The multiplier of the RPS program needs to be readjusted. According to the current RPS program, offshore wind, tidal energy and fuel cells have a higher multiplier values than others as indicated in Table 2.2. There are practical and legal reasons why the multiplier should be readjusted.

In the case of tidal power, with its relatively higher multiplier and reasons of economy of scale, there have been plans to build tidal power plants in four areas in Korea including the Sihwa Lake, Incheon Bay, Ganghwa Bay, and Garorim Bay. As the world's largest ocean project, the Sihwa Lake tidal power station with a generation capacity of 254 MW was already commissioned in August 2011.[56] However, the tidal power construction plans in all other areas received strong opposition from environmental NGOs and local communities. The Incheon tidal power station was planned to be the largest tidal power plant in the world with a generation capacity of 1,320 MW along with 44 water turbines rated at 30 MW each. The construction and development costs were expected to be KRW 3.9 trillion. This tidal power plant construction was opposed by local communities and environmental NGOs due to mistrust of prior environmental review and business feasibility.[57] The Garorim Bay tidal power project was a KRW 1 trillion project of KOWEPO, POSCO Power, Daewoo Engineering and Construction, and Lotte Engineering & Construction Co. Ltd. It was planned to build a power generation plant with a capacity of 520 MW with a 2 km long concrete tide embankment at the mouth of the bay. The project was strongly opposed by environmental NGOs due to severe negative impacts on the environment of the bay.[58] The

---

55. *Ibid.*, 43.
56. A tide embankment at the Sihwa Lake was constructed in 1994 for flood mitigation and agricultural purposes. After the tide embankment was built, pollution built up in the newly created Sihwa Lake reservoir, which caused much environmental concern. Compared to other tidal power plants, the tidal power plant at Sihwa Lake was not strongly opposed against by environmental NGOs and the civil society in the beginning of the construction since it was being built on the already existing tide embankment. However, after the operation of the tidal power plant, there have been complaints arising from environmental NGOs and local fishermen due to the phenomenon of green algae and appearance of jellyfish, which disturbed fishing activities. Tae-Seong Hahm, *A Legal Study on Sustainability and Marine Environment Preservation*, 34 Environmental Law Review 57, 67 (2012).
57. Local communities and environmental NGOs regarded that environmental consultation processes were not properly conducted. *Ibid.*, 67.
58. Citing research reports by the Ministry of Maritime Affairs and Fisheries, a well-known Korean environmental NGO argues that environmental values of the Garorim Bay were rated as the highest among all the coastal areas of the country. The construction of tide embankment in the

Ministry of Environment in Korea expressed its environmental concerns after rejecting the Environmental Impact Assessment III report on the project in April, 2012.[59] As such, all of the construction plans, as of July 2013, are on hold due to social, environmental and political reasons.

These series of events show that the current RPS program needs to carefully readjust the multiplier value of certain NRE source such as tidal power. Furthermore, the current RPS program needs to limit the size of the NRE facilities through giving smaller or no multiplier values if the project exceeds a certain size. For example, the Enforcement Decree of the Act on the Promotion of the Development, Use and Diffusion of New and Renewable Energy can exclude the tidal power plants exceeding a certain size if that plant diminishes the qualities of environmental conditions in local areas.[60]

Additionally, the current RPS program needs to harmonize with other laws. All abovementioned controversial tidal power plant constructions had problems in the decision making processes including consultations with stakeholders such as local residents. Consultations with local residents were not made at the early stages of project developments. Furthermore, consultations were simply regarded as processes of informing the relevant local residents about the impacts of projects. This should be changed in order to expand NRE in the current RPS program. Thus, the RPS program should be well related with the Environmental Impact Assessment revised and implemented after July, 2012.[61] This way will increase the public acceptance of NRE projects under the current RPS program, resulting in much lesser social and environmental conflicts due to the construction of NRE facilities.

The RPS program can also create conflicts with existing laws for protecting the natural environment in Korea. Resolving these conflicts can also contribute to increasing the successful implementation of the RPS program as well as protecting Korea's nationwide ecological corridors, such as the Baekdu Daegan mountain range and coastal zones. For example, the plan of constructing a tidal power plant at Inchon bay included the tidal flat area of Jangbong Island, which was designated as a wetland protection area in December 2003. In order to build the tidal power plant, the

---

area was considered by this NGO as an activity that put important species into the danger of extinction. The bay was known as an area for one of only two habitats of the Spotted Seal Phoca largha protected under the Endangered Species Category II designated by the Ministry of Environment, and an area for an important number of other migratory waterbird species in South Korea. Korean Federation for Environmental Movement(KFEM), *Tidal Power Projects of South Korea and Their Impacts on Tidal Flat Conservation*, http://koreawetlands.blogspot.kr/2009/11/impacts-of-tidal-power-projects-of-s.html (accessed 13 Jul. 2013).

59. MOE gave the following reasons for its rejection of the EIA report: lack of review of the change of erosion and deposition over time with the construction of tide embankment, lack of consideration of water pollution, lack of research and protection measures of endangered species. Hahm, *supra* n. 54, at 69.
60. Hahm, *supra* n. 54, at 78. *See also* Kwon, *supra* n. 12, at 158.
61. Revised Environmental Impact Assessment Law combined prior environmental review and environmental impact assessment under one single law. It regulates environmental impact assessment under three categories such as Strategic Environmental Impact Assessment (SEIA), Environmental Impact Assessment (EIA), and small scale Environmental Impact Assessment. Kwon, *supra* n. 12, at 79.

designation of the protection area would need to be cancelled. Current laws provide very vague or widely inclusive regulations in relation to the cancellation of the designation of wetland protection areas, which likely result in canceling such designation by the competent administrative agency with its arbitrary interpretation and standards.[62] This very same situation can occur with other NRE sources such as wind power, in which areas in the Baekdu Daegan mountain range are considered to be suitable sites for wind power facilities due to the presence of strong winds. The Ministry of Environment has been strongly criticized by other relevant ministries and power generators about regulating construction of wind power facilities in the protected areas. Recently, government agencies such as the Ministry of Environment and the Ministry of Trade, Industry and Energy discussed to deregulate the conditions of locations for wind power facilities. The Ministry of Environment announced that it will prepare guidelines for the location of onshore wind power facilities that can harmonize with the environment by the end of 2013.[63]

As shown in the cases of tidal power and wind power, the two aims conflict. In a way, both are designed to protect the environment. However, protecting a certain part of environmental qualities hinders protecting other quality of the environment, in which the dilemma exits. Thus, it needs a careful examination to meet the targets for both purposes, one of protecting the natural environment and the other of generating energy from renewable energy sources. For ensuring the long-term sustainability of Korea, it is necessary not to allow the cancellation of the designation of the protection areas once they are designated for the purpose of ecological protection. However, in order to solve the dilemma between the protection of the natural environment and the generation of energy from renewable energy sources, there is, again a necessity arising to reform or readjust of the current RPS program in Korea.

Scholars, thus, suggest that the multiplier of tidal power should be readjusted to be one third or one fourth of the multipliers of wind power or PV.[64] In the case of

---

62. Section 1 of Art. 10 of the Wetland Conservation Act states, 'The Minister of Environment, the Minister of Land, Transport and Maritime Affairs or the Mayor/Do governor may cancel the designation of or reduce the size of the wetlands protection area, etc. which fall under the inevitable cases of if it is unavoidable for the public benefit or military purpose as prescribed by Presidential Decree, or which have lost value or have become unnecessary to conserve as the wetlands protection area, etc. because of natural disasters or other causes.' And s. 3 of Art. 25 of the Conservation and Management of Marine Ecosystems Act states, 'The Minister of Land, Transport and Maritime Affairs may alter or cancel the designation of protected marine areas, when such areas lose the value of protected marine areas under paragraph (1) or do not need to be conserved, due to military purposes, natural disasters or other grounds'. *Ibid.*, 81.
63. Ministry of Environment, *Press Release in Responding to the News Article of Chosun Daily Newspaper*, http://www.me.go.kr (accessed 19 Jul. 2013). The Ministry of Environment recently conducted on-site survey on 14 onshore wind power farms to examine feasibility of locations. In the results of the prior environmental review, the Ministry examined whether construction of wind power facilities significantly damages geographic features, destroys ecosystems, or affects local residents' health. Particularly, the Ministry concluded that some of wind power facilities are very likely to damage the environment in the process of building access and power transmission facilities. Energy Korea, *Guideline for Onshore Wind Power Farms Seen to be Worked Out by End of This Year*, http://energy.korea.com/archives/55251 (accessed 23 Jul. 2013).
64. Hahm, *supra* note 54, at 82.

onshore wind power that conflicts with the protection of natural environment, the multiplier should be re-categorized according to the land type. If the onshore wind power farm is located in ecologically sensitive areas, the multiplier should be less than the current multiplier (1.0). If it is in less ecologically sensitive areas, the current multiplier can be maintained.

The solution for reducing the abovementioned conflicts requires not only legal change but also institutional change. The absence of a mechanism to coordinate conflicts among different stakeholders' interests on the successful implementation of the RPS program can be an important factor for successful RPS program implementation. Currently, there is no institution to coordinate the resolution of the abovementioned conflicts. Thus, one way of resolving those conflicts can be creating a mechanism to coordinate the different demands and responsibilities of relevant ministries and other stakeholders.[65]

The readjustment or reform of the multiplier of the current RPS program is also applicable to the solar carve-out program. Some suggest readjusting the multipliers for cases not using buildings and existing facilities. For example, the multiplier for five areas such as dry field can be increased if the way of facilitating PV does not degrade the quality of land. Additionally, areas such as abandoned salt fields where there would be no harm done to the quality of the land and the capacity for solar power is over 30 kW can have a greater multiplier than the current multiplier (1.0).[66] Others suggest that the Act on the Promotion of the Development, Use and Diffusion of New and Renewable Energy needs to be revised since the fact that the Enactment Decree of the Act only includes solar carve-out provisions and is against the provision of 'new and renewable energy necessary for the balanced use and distribution' in the Act. In order to designate a specific new and energy source such as PV in the Enactment Decree, the provision of the Act should include 'new and renewable energy necessary for the balanced use and distribution and with high degree of distribution' since PV has a higher degree of distribution compared to other NRE sources.[67]

*[2]     Eligible Renewables*

Not including non-electricity generation technologies in the RPS program is one of its shortcomings for expanding renewable energy. The current RPS program of Korea needs to diversify renewable energy sources and to carefully reconsider which energy sources can be included in the program. Originally, geothermal energy was included into planned RPS program and was announced to have the multiplier of 2.0 at the public hearing for establishing the RPS program. However, the final RPS program did

---

65. The absence of coordinating institutions was one of main reasons of an early walkout of the FIT program. Min Gyo Koo, *South Korea's Feed-in Tariff Program and Its Implications for New Industrial Policy: Policy Design, Implementation, and Learning*, 22 Korean Public Administration Review 1, 17 (2013).
66. Sun-Jin Yun, 'The Institutional Improvement of RPS: Focusing on Solar Photovoltaics', Presented at the Second Policy Forum for the Promotion for Use of Solar Photovoltaics, http://eco.or.kr/11968/ (accessed 20 Jul. 2013).
67. Yi, Yoon and Park, *supra* n. 7, at 302.

not include geothermal energy. Korea has sufficient geothermal power potential, and there is an Enhanced Geothermal System (EGS) technology to develop geothermal energy. Thus, critics argue that the current RPS program fails to make a balance between different NRE sources since the Korean government includes and provides the highest multiplier value to fuel cells that takes time for technological development and commercialization and includes IGCC, which is not a renewable energy source.[68]

### [3]   Dual Track System of FIT and RPS

The current RPS program needs to be implemented with other NRE support or climate change measures such as the FIT program and the Emission Trading Scheme (ETS). The main reason of the shift from FIT to RPS was not because the policy effectiveness of FIT was not sufficient but because the financial burden on the government had significantly increased. Thus, in order to fulfill the two main purposes of increasing deployment of NRE and reducing financial burden, generators can be regulated under two different programs. For example, large scale generators can be regulated under the RPS program, while small scale generators can be regulated or supported under the FIT program.[69]

### [4]   Linking the RPS and ETS

The RPS program also needs to resolve problems associated with the complaints from targeted generators about dual obligations under the RPS program as well as the ETS scheduled to begin in 2015.[70] Mutual recognition of REC and emission rights in the market is one way to solve the problem. In other words, REC can be used for the purpose of carbon emission reduction targets while emission rights can be used for satisfying RPS obligations. This can be quite a flexible mechanism. However, in cases where the market prices of emission rights are low, generators tend to easily purchase

---

68. Inho Choi, *A Study on Regulatory and Institutional Obstacles to the Promotion of Development of Deep Underground Geothermal Energy in the Republic of Korea*, 33 Environment Law Review, 254, 299 (2012).
69. Hahm, *supra* n. 54, at 82. *See also* Hyung-Jun Hwang, 'Issues related with the RPS Program: Focusing on the Britain's Renewable Obligation(RO) Program', in Hong Sik Cho and others (eds), *Energy Law and Policy in the Age of Climate Change* (Parkyoungsa, 2013) 221. Energy Act 2008 of the UK introduced a FIT program to incentivize small scale electricity generators from renewable sources (less than 5 MW). Matthew Leach and Sandip Deshmukh, 'Sustainable Energy Law and Policy', in Karen E. Makuch and Ricardo Pereira (eds), *Environmental and Energy Law* 136 (Wiley-Blackwell 2012). While the national government does not adopt this idea yet in the current RPS program, the Seoul metropolitan government announced its plan to implement the FIT program for small scale generators.
70. Act on the Allocation and Trade of Greenhouse Gas Emissions Rights was passed in the National Assembly in May 2012. Three phases are outlined for the scheme. The first phase of the trading scheme is planned to begin in 2015 and end in 2017. It covers companies that emit 125,000 metric tons or more of carbon dioxide a year and factories, buildings and livestock farms that produce at least 25,000 tons of the gas annually. The second phase starts from 2018 and end in 2020. The third phase, unlike previous phases, is expected to be a five-year term. In the initial phase, the government plans to offer over 95% of all carbon credits for free and will auction the remainder. IEA, *supra* n. 3, at 12–13.

emission rights rather than try to satisfy their RPS obligations, which ultimately results in failing to satisfy the target of the deployment of NRE. On the other hand, while REC can be traded at the emission trading market, emission rights cannot be traded at REC markets. This can fulfill RPS targets for generators by providing flexibility. However, this is also related to the difficulty of deciding the relative price between REC and emission right and allowing $CO_2$ emission for the amount of generating energy from NRE sources.[71] In order to solve these problems, policy makers need to discuss this issue before implementing the ETS in 2015.

## §2.06 CONCLUSION

Renewable energy has been pursued and expanded as an alternative energy source in tackling climate change and in ensuring stable energy provision. However, there are limits in expanding NRE since environmental conditions and generation costs are important factors for the deployment and expansion of NRE in Korea. Renewable energy can be highly influenced by the national land scale and the environmental conditions such as hydropower with water resource, solar energy with gross area, wind power with wind condition and gross area, bio energy with forest and grain resources, etc. Furthermore, construction of NRE facilities involves the process of persuading local communities in which those facilities are built. Thus, it is crucial to carefully consider these limitations and strategically choose policy tools to expand NRE.

It is too early to assert whether the current RPS program has failed since it has been only a year since its implementation. However, diverse evidence suggests that the current RPS program needs to be improved through readjustment of multipliers, limiting generation capacities, excluding environmentally problematic energy sources, harmonizing with other laws and policy tools, creating a mechanism for resolving conflicts among different stakeholders, and including diverse renewable resources rather than investing more on fossil-fuel based energy sources. Additionally the current RPS program needs to consider how to incorporate the program with FIT and ETS. Policymakers are expected to encounter complaints from mandatory suppliers who have to pay penalties for not fulfilling requirements and demands to reduce their mandatory obligations. As seen from the cases in other countries, if the RPS program is weakened due to the political bargaining process, there will be no real growth of renewable energy. To prevent the political bargaining process from overwhelming the existing program, all discussed suggestions in this chapter should be carefully studied and adopted in a suitable way.

---

71. Hwang, *supra* n. 67, at 226.

CHAPTER 3
# A More Sustainable Way to Promote PV: Transformations from FIT to FIT/FIT Tendering Schemes in Taiwan and France

*Anton Ming-Zhi Gao*[*]

The adoption of a feed-in tariff (FIT) mechanism can be a suitable method to promote the installation of renewable energy technologies. However, this very popular tool can also be the victim of its own success: In some cases, the rates set for FIT schemes can lead to an unsustainable technology-installation boom within a very short period. For example, the recent explosive development of photovoltaic (PV) modules led to hikes in electricity prices in many countries that adopted FIT schemes. Almost simultaneously, Taiwan and France experienced unexpected rushes to develop PV installations. This forced both governments to alter the designs of their FIT policy mechanisms. They adopted tendering schemes to rein in the rates for PV installation while continuing to facilitate the industry's development. These modifications appear to have met their design purposes. Numerous countries around the world are set to incorporate tendering schemes. However, questions related to the complexity of these schemes, long-term security for investors, and even the tendering scheme designs have been raised. This study examines the structure and implementation of tendering schemes in Taiwan and France. Furthermore, this study assesses these schemes with respect to the suitability of promoting PV installations. This study attempts to determine these tendering mechanisms' degrees of stability and success. It also offers suggestions for ways to reform tendering schemes if they are fundamentally flawed or unsuitable.

---

[*] This article was funded by the National Science Council, Taiwan, Project Number: NSC101-2410-H-007-024-MY2.

## §3.01 INTRODUCTION

Most countries have become very active and keen to adopt both 'mitigation' and 'adaptation' measures and strategies to address the global warming crisis. The main goal of mitigation strategies is to promote the development of renewable energy and renewable electricity (RE). Both of these forms of energy can help improve national economies and enhance global competitiveness. In particular, wind energy and photovoltaic energy (PV) are becoming primary energy sources. They demonstrate significant growth potential. This article will focus on PV, and describe recent and dramatically reformed legal promotion schemes being applied in Taiwan and France.

With respect to the development of PV, Taiwan and France both possess great sunshine resources and good geological locations. They also possess industrial advantages that provide foundations for enduring policy supports. With respect to promotion regimes, both countries have adopted effective mainstream regimes that imposed feed-in tariffs (FIT) during their first phases and have attained significant achievements. For instance, during 2009, the PV growth rate in France ranked fifth in Europe.[1] In 2010, the PV growth rate in France equaled the PV growth rate in the U.S.[2] In 2010, Taiwan's scheduled PV development target was 75 MW. However, more than 10 times the scheduled number of cases flooded in. The actual number of cases equaled the scheduled development target for 2015–2020. This PV proliferation or PV 'boom' phenomenon also attracted governmental attention/concerns. Therefore, further reforms were introduced. For instance, both Taiwan and France launched FIT reforms in 2010[3] and 2011.[4] These types of reforms also triggered chain reactions in other countries that adopted FIT schemes.

The most interesting points related to FIT reform in Taiwan and France involve 'coincident timing' that occurred during the reform of FIT regimes, in spite of the geological locations and language differences between these two countries. Firstly, in December 2010, both countries adopted controversial 'preliminary' measures to address the PV boom. For example, France adopted a decree to cease all applications for cases that involved over 3 kW PV installations for three months beginning on 9 December 2010.[5] Taiwan changed the application rule of FIT rate and retrospectively

---

1. *See* Baromètre Photovoltaïque, Photovoltaic Barometer, p. 131, *available online* at http://www.eurobserv-er.org/pdf/baro196.pdf.
2. *See* Alliance for Renewable Energy, FITs: Higher Biogas – Lower PV Tariffs Coming *available online* at http://www.allianceforrenewableenergy.org/2011/03/flexible-french-fits-higher-biogas-lower-pv-tariffs-coming.html.
3. Arrêté du 31 août 2010 fixant les conditions d'achat de l'électricité produite par les installations utilisant l'énergie radiative du soleil telles que visées au 3 de l'article 2 du décret n 2000-1196 du 6 décembre 2000.
4. *See* Reuters Africa, Factbox-Global Support for Wind and Solar Power, 2011, *available online* at http://af.reuters.com/article/energyOilNews/idAFLDE73D23920110428; Climate Change Report, Recent Significant Changes in Photovoltaic Energy Regulation in France, 20 Apr. 2011, *available online* at http://www.klgates.com/recent-significant-changes-in-photovoltaic-energy-regulation-in-france-04-13-2011/.
5. Décret n 2010-1510 du 9 décembre 2010 suspendant l'obligation d'achat de l'électricité produite par certaines installations utilisant l'énergie radiative du soleil.

Chapter 3: FIT and Tendering Scheme in Taiwan and France          §3.01

applied it to applications made during that year on 17 December 2010.[6] Subsequently, in an attempt to further reform the FIT scheme, in March 2011, France and Taiwan introduced combo schemes that involved both tendering and FIT. France issued three administrative orders and introduced a simple tendering scheme and a complex tendering to address 100~250 kW rooftop and >250 kW rooftop/all ground PV, respectively. However, <100 kW rooftop PV remains subject to the FIT scheme.[7] Taiwan initiated its combo scheme on 17 March 2011.[8] Installations are divided into four categories, based on equipment type, installed capacity, and whether installations are self-owned. With the exception of Category 1 (capacity between 1 kW–10 kW, rooftop installation self-owned), which will remain subject to FIT, all applications from the remaining three categories[9] are subject to the tendering scheme.

France and Taiwan appear to be leading the PV reform boom. Tendering schemes originally applied to wind electricity power[10] are being applied to PV sectors around the world (e.g., Australia,[11] California in the U.S.,[12] South Africa,[13] and India).[14] Thus, it is apparent that France and Taiwan have become the leaders in PV subsidy reform after the previous phase when German FIT scheme prevailed.

The purpose of this study is to examine the practical implementation of PV tendering systems in Taiwan and France. Particularly, a fundamental question must be asked: Are tendering systems better than traditional FIT schemes for the promotion of sustainability in the PV industry? Alternatively, as indicated by the literature, the adoption of tendering systems cannot provide 'long-term investment security'[15] for PV industries in comparison with FIT schemes. Hence, are adjustments to the tendering design required?

---

6. Ordinance of FIT Rate Schedule of 2010.
7. For an introduction to the tendering scheme in France, *see*, Frass-Ehrfeld, Clarisse, Renewable Energy Sources: A Chance to Combat Climate Change 370 (2009).
8. PV Tendering Ordinance of Phase I of 2011.
9. (1) Rooftop, with an installed capacity between 1 kW–10 kW and not owned by the building owner; (2) rooftop installation with an installed capacity superior to 10 kW; (3) ground-type installation, with an installed capacity superior to 1 kW that fulfills the 'Land Use Regulations'. See, Art. 3 of PV Tendering Ordinance of Phase I of 2011.
10. Such as in Ireland, France, U.K., China, Denmark, and so on. *See*, e.g., Paul Gipe, Wind Power: Renewable Energy for Home, Farm, and Business 212 (2004); Frass-Ehrfeld, Clarisse Fräss-Ehrfeld, Renewable Energy Sources: A Chance to Combat Climate Change 265 (2009).
11. *See*, e.g., Thiess-Silex Solar Consortium confirmed as major bidder to build large scale solar photovoltaic (PV) power station in Australia, 10 May 2010, *available online* at http://www.aeol.com.au/databases/news/thiess_silex_solar_sonsortium_confirmed.htm.
12. *See*, e.g., Winning bids for California's RAM come in under USD 0.089/kWh, *available online* at http://www.solarserver.com/solar-magazine/solar-news/current/2012/kw14/winning-bids-for-californias-ram-come-in-under-usd-0089kwh.html.
13. *See*, e.g., Solar Capital also successful in South African PV bid, 12 Dec. 2011, *available online* at http://www.pv-magazine.com/news/details/beitrag/solar-capital-also-successful-in-south-african-pv-bid_100005222/#axzz1uFDSFBY1.
14. *See*, e.g., India's Karnataka invites solar bids worth 80 MW, PV-Magazine, *available online* at http://www.pv-magazine.com/news/details/beitrag/indias-karnataka-invites-solar-bids-worth-80-mw_100003925/.
15. 'This [tendering] method gives less long-time investment security than the feed-in tariff does'. *See* e.g., Manfred Stiebler, Wind energy systems for electric power generation 9 (2008).

To answer the aforementioned research questions, the authors collected information and documentation related to tendering schemes used in Taiwan and France. Then, they compared and analyzed that information. Second, this study examines the practical implementation of tendering systems in both countries and analyzes contributions made to the development of PV industrial installations. Finally, this study offers some suggestions for reforms based on comparisons and contrasts of PV development in FIT design and tendering systems in Taiwan and France.

## §3.02 TENDERING SCHEMES IN FRANCE

### [A] Background: The Adoption of FIT and Tendering Schemes in 2000

To address the needs of both RE promotion and electricity liberalization, France passed legislation focused on electricity liberalization in 2000.[16] On the one hand, based on Germany's successful FIT efforts, France introduced a FIT scheme. RE installers can enjoy long-term (12–15–20 years) contracts and electricity can be purchased at favorable rates. For instance, according to the 2002 Decree for PV FIT,[17] the PV rate for continental France is 15.25 Ct/kWh for 20 years. In contrast, the PV rate for overseas France is 30.5 Ct/kWh for 20 years.[18]

On the other hand, a tendering scheme was also adopted to promote RE.[19] Under this scheme, the government would fix the development cap and open calls for tenders from potential investors to win/earn power purchase contracts.[20] Originally, during the 2000s, this tendering scheme was mainly applied to onshore and offshore wind[21] and

---

16. Article 10, Loi n 2000-108 du 10 février 2000 relative à la modernisation et au développement du service public de l'électricité.
17. Annexe 1 Tarifs mentionnés a l'article 5 de l'arrêté, Arrêté du 13 mars 2002.
18. Article 5, Arrêté du 13 mars 2002 fixant les conditions d'achat de l'électricité produite par les installations utilisant l'énergie radiative dusoleil telles que visées au 3o de l'article 2 du décret no 2000-1196 du 6 décembre 2000, *available online* at http://admi.net/jo/20020314/ECOI0200002A.html (last retriever 2 Aug. 2013).
19. Article 8, Loi n 2000-108 du 10 février 2000 relative à la modernisation et au développement du service public de l'électricité.
20. Steven Ferrey, Unlocking the global warning toolbox: Key choices for carbon restriction and sequestration, 247–248 (2010).
21. To learn more about the tendering specifications for onshore wind, *see*, Cahier des charges de l'appel d'offres n 332689-2010-FR portant sur des installations éoliennes terrestres de production d'électricité en Corse, Guadeloupe, Guyane, Martinique, à La Réunion, à Saint-Barthélemy et à Saint-Martin, *available online* at http://www.cre.fr/documents/appels-d-offres/appel-d-offres-portant-sur-des-installations-eoliennes-terrestres-de-production-d-electricite-en-corse-et-outre-mer/cahier-des-charges. To learn more about the tendering specifications for offshore wind, *see*, Cahier des charges de l'appel d'offres n 2011/S 126-208873 portant sur des installations éoliennes de production d'électricité en mer en France métropolitaine, *available online* at http://www.cre.fr/documents/appels-d-offres/appel-d-offres-portant-sur-des-installations-eoliennes-de-production-d-electricite-en-mer-en-france-metropolitaine/cahier-des-charges-version-rectifiee-du-21-11-2011-appel-d-offres-portant-sur-des-installations-eoliennes-de-production-d-electricite-en-mer-en-france-metropolitaine.

biomass electricity.[22] However, due to concerns related to the aforementioned PV boom, France began to apply the tendering scheme to the PV sector in March 2011.

### [B]   Combo Tendering Scheme and FIT in 2011 and 2013

The French government adopted the combo scheme on 4 March 2011. Under this scheme, ≤100 kW rooftop PV is subject to FIT. In contrast, all ground PV and > 100 kW rooftop PV is subject to the tendering scheme.

The tendering scheme appears in two forms. First, the simple tendering scheme is applied to ' > 100 kW ~ ≤ 250 kW' rooftop PV. The main factor required to win a bid is determined by the bid's tendering price.[23] Second, the complex tendering scheme is applied to all ground (< 12 MW) PV and > 250 kW rooftop PV. The complexity developed because of complex and multiple winning bid factors that included bidding prices, environmental effects, project feasibility, contributions to research and development (R&D), after-life recycling duties, and so on.[24]

These schemes intended to run until 2014. Yet, to reflect the results of five applications of simple tendering and one application of complex tendering, the French government decided to cancel the remaining two applications of simple tendering. In 2013, it instituted new and similar (but slightly different) tendering schemes. This new scheme includes two tendering schemes. It will remain valid until 2016. At a glance, the main difference between these schemes is that the new simple tendering scheme is more complex than the older scheme. The new complex tendering scheme has been simplified. However, because the tendering results of these new schemes will only become available after this article's publication, this article will only elaborate on the designs of these new schemes.

This article will further elaborate on these two tendering schemes and discuss the results of the 2011 tendering scheme. Because the aforementioned administrative decrees merely provide rough information related to the simple and complex tendering schemes, details of the tendering specifications are further elaborated in the following four key documents, which will be discussed in this article:

---

22. CRE, Cahier des charges de l'appel d'offres n 2010/S 143-220129 portant sur des installations de production d'électricité à partir de biomasse, *available online* at: http://www.cre.fr/documents/appels-d-offres/appel-d-offres-portant-sur-des-installations-de-production-d-electricite-a-partir-de-biomasse/cahier-des-charges.
23. *See* Énergie Solaire Photovoltaïque: Le Nouveau Dispositif, *available online* at http://www.renewablesb2b.com/ahk_tunisia/fr/portal/solar/news/show/7bcd5d000d7b7f0a. (last retrieved 2 Aug. 2013).
24. *See* Olivia Lê Horovitz et al., Recent Significant Changes in Photovoltaic Energy Regulation in France, 19 Apr. 2011, *available online* at http://www.klgates.com/recent-significant-changes-in-photovoltaic-energy-regulation-in-france-04-13-2011/ (last retrieved 2 Aug. 2013). *See* Comité de Liaison Énergies Renouvelables (CLER), Nouveaux tarifs du solaire photovoltaïque, *03-22-2011*, *available online* at http://www.cler.org/info/spip.php?article9291 (last retrieved 2 Aug. 2013).

(1) 2011 Simple Tendering Specification[25] ('2011 Simple Spec').
(2) 2013 Simple Tendering Specification[26] ('2013 Simple Spec').
(3) 2011 Complex Tendering Specification[27] ('2011 Complex Spec').
(4) 2013 Complex Tendering Specification[28] ('2013 Complex Spec').

### [C] Simple Tendering Scheme in 2011 and 2013

#### [1] Policy Targets

To promote the use of renewable energy and PV, France government set both renewable 'energy' and PV targets for 2020: Renewable energy for energy consumption is scheduled to reach 23%, which is equal to 20 million tons of oil equivalent (Mtoe) out of total annual production of renewable energy. The PV development target was set at 5,400 MW.[29] These targets have been maintained, and were repeated in the 2013 Simple Spec.[30]

Although PV installers and investors usually oppose the tendering scheme and support the FIT scheme,[31] the government still hopes to achieve its targets *gradually* under the tendering scheme, rather than *suddenly* under the FIT scheme.

---

25. CRE, Cahier des charges de l'appel d'offres portant sur la réalisation et l'exploitation d'installations photovoltaïques sur bâtiment de puissance crête comprise entre 100 et 250 kW, *available online* at http://www.cre.fr/documents/appels-d-offres/appel-d-offres-portant-sur-des-installations-photovoltaiques-sur-batiment-de-puissance-crete-comprise-entre-100-et-250-kw/cahier-des-charges-de-l-appel-d-offres-portant-sur-des-installations-photovoltaiques-sur-ba timent-de-puissance-crete-comprise-entre-100-et-250-kw (last retrieved 2 Aug. 2013).
26. Cahier des charges de l'appel d'offres portant sur la réalisation et l'exploitation d'installations photovoltaïques sur bâtiment de puissance crête comprise entre 100 et 250 kW, *available online* at http://www.cre.fr/documents/appels-d-offres/appel-d-offres-portant-sur-des-installations-photovoltaiques-sur-batiment-de-puissance-crete-comprise-entre-100-et-250-kw2/cahier-des-charges-ao-150-250-22032013.
27. Appel d'offres portant sur la réalisation et l'exploitation d'installations de production d'électricité à partir de l'énergie solaire d'une puissance supérieure à 250 kWc, Cahier des charges (500,92 ko), *available online* at http://www.cre.fr/documents/appels-d-offres/appel-d-offres-portant-sur-la-realisation-et-l-exploitation-d-installations-de-production-d-electricite-a-partir-de-l-energie-solaire-d-une-puissance-superieure-a-250-kwc/cahier-des-charges
28. Cahier des charges de l'appel d'offres portant sur la réalisation et l'exploitation d'installations de production d'électricité à partir de l'énergie solaire d'une puissance supérieure à 250 kWc, available at: http://www.cre.fr/documents/appels-d-offres/appel-d-offres-portant-sur-la-reali sation-et-l-exploitation-d-installations-de-production-d-electricite-a-partir-de-l-energie-solaire-d-u ne-puissance-superieure-a-250-kwc2/telecharger-le-cahier-des-charges.
29. 2011 Simple Spec, at 1.
30. 2013 Simple Spec, at 1.
31. ENERPLAN, Enerplan demande deux mesures pour relancer la filière de l'énergie solaire en France et créer des emplois, 25 janvier 2012, *available online* at http://www.photovoltaique. info/IMG/pdf/120125_cp_enerplan_deux_mesures_pour_relancer_le_solaire.pdf (last retrieved 2 Aug. 2013).

## [2] Tendering Targets and Caps

In the 2011 Simple Spec, France set a target of 300 MW for ' > 100 kW ~ ≤250 kW' rooftop PV to be achieved between mid-2011 and 2014.[32] The seven-phase tendering schedule is listed in Table 3.1.[33] In practice, the tendering scheme has already completed its fifth phase. The sixth and seventh phases were cancelled and integrated into the 2013 Simple Spec.

*Table 3.1  Tendering Schedule and Capacity under the 2011 Simple Tendering Scheme*

| Phase | Tendering Capacity | Period: Call for Tenders |
|---|---|---|
| 1 | 120 MW | 2011/8/1-2012/1/20 |
| 2 | 30 MW | 2012/1/21-3/31 |
| 3 | 30 MW | 2012/4/1-6/30 |
| 4 | 30 MW | 2012/7/1-9/30 |
| 5 | 30 MW | 2012/10/1-12/31 |
| 6 | 30 MW | 2013/1/1-3/31 |
| 7 | 30 MW | 2013/4/1-6/30 |

*Source*: 3.2 Périodes de candidature, 2011 Simple Spec.

In the 2013 Simple Spec, a new three-phase schedule will be introduced. The schedule should reach 120 MW in 2016, which will be divided into three phases of 40 MW each. The three phases will be 1 July, 1-31 October 2013; 1 November 2013-28 February 2014; and 1 March-30 June 2014, respectively.[34]

## [3] Eligibility

The 2011 and 2013 Simple Specs include the following eligibility requirements.

### [a] Installation Capacity

Tendering should ensure that installation capacity is ' > 100 kW ~ ≤250 kW' installation capacity.[35] In addition, to avoid situations in which one legal entity submits and is awarded too many cases, a ceiling limit of 250 kW has been set for applicants and subsidiaries located within the same buildings or within a 500 meters range.[36]

---

32. 2011 Simple Spec, at 1.
33. 3.2 Périodes de candidature, 2011 Simple Spec.
34. 2.1, 2013 Simple Spec.
35. 3.1 Caractéristiques des installations, 2011 Simple Spec. *See also*, 2013 Simple Spec, 3.1.
36. 2011 Simple Spec, 3.1; 2013 Simple Spec, 3.1.

*[b] Capacity Requirements for Each Case*

The Capacity Requirements for each case are similar to the Installation Capacities. They can be defined as > 100 kW ~ ≤250 kW.[37]

*[c] Other Criteria*

The 2011 Simple Spec provides additional criteria required for applications:

(1) Simple BIPV: Each installation should be a rooftop system and comply with the criteria outlined in the *simplified building integrated PV (BIPV) requirement* in Annex 2 of the Decree of 4 March 2011. An installation that covers part or all of a parking lot (i.e., the so-called '*shading park lot*'), is strictly prohibited.[38]
(2) Certificate requirement: Each manufacturer of related installations (including modules, inverters, electrical equipment, and so on) must possess related certificates, such as ISO 9001, ISO 14001, or other equivalent certificates issued by the Certificateur Accrédité par le Comité Français d'accréditation (COFRAC) or other similar European association.[39]
(3) Grid Connection Management: Each applicant must demonstrate its financial participation in tendering grid connections. At the very least, the possibility of indirect connections should be secured.[40]
(4) Local government's participation in the tendering process: The local government is allowed to participate in the tendering process.[41]
(5) Recycling duties: Each applicant must demonstrate that he/she has arranged with a professional recycling organization to recycle related film PV modules once those modules cease operation. Each applicant agrees to pay recycling fees charged by authorities, if required.[42]
(6) Financial guarantee requirements and loan offers: Each applicant must certify that it is financially capable to complete the project (a deposit of *EUR 0.6/W*) is required for each installation). Each applicant must successfully obtain a loan offer from a related bank.[43]

In the 2013 Simple Spec, most requirements were retained, with the exception of the grid connection management requirement. Grid connection occurs on a case-by-case basis. Therefore, in this new Spec, each applicant must complete and commercially operate the installation within two months after grid connection is completed.[44]

---

37. 2011 Simple Spec, 3.1 Caractéristiques des installations. See also, 2013 Simple Spec, 3.1.
38. 2011 Simple Spec, 3.1. See also, 2013 Simple Spec, 3.1.
39. 2011 Simple Spec, 3.1. See also, 2013 Simple Spec, 3.1.
40. 2011 Simple Spec, 5.1.
41. 2011 Simple Spec, at p. 1.
42. 2011 Simple Spec, 3.4. See also, 2013 Simple Spec, 3.3.
43. 2013 Simple Spec, 3.1 and Annex 3.
44. 2013 Simple Spec. 3, 3.2.

In addition, the 2013 Simple Spec provides additional extra requirements:[45]

(1) Each installation must be *new* or *semi-new*. With respect to semi-new installations, each installation will be considered new if it was not in service at the time each application was filed.
(2) Securing related buildings' control rights: Each applicant must secure installation sites for 20 years (e.g., sign lease agreements).
(3) Planning and building permits: Each applicant must provide planning permissions and copies of building permits or certificates at the time each application is filed.

*[4]    The Ratio of Rooftops and Ground PV*

Because only rooftop PV is qualified under the simple tendering scheme, this question is not relevant.

*[5]    Administrative Procedures for Tendering*

*[a]    Submission of Related Documents*

The red tape requirements that affect the 2011 Simple Spec are relatively bothersome. Applicants must submit many documents and complete many forms. These include:

- Annex I document (application form):[46] This document must contain applicant's basic information (e.g., applicant's name, contact address, and so on) and a very comprehensive description of the project details (e.g., manufacturer, manufacturer's location, PV module and inverter brands, and so on) and the amount of estimated investments (amount of equity, amount of debt, amount of investment subsidies, and so on).
- Annex II requires the provision of copies of planning permissions for the installation, the ISO or related certificates for PV-related installations, a certificate to ensure that 'financial guarantee requirements and loan offers' have been provided by bank (the format is provided in Annex 3[47]), and the provision of a proposal that does not exceed six pages.[48]

---
45. 2013 Simple Spec, 3.1.
46. 2011 Simple Spec, Annexe 1: Copie du formulaire de candidature en ligne.
47. 2011 Simple Spec, Annexe 3: Modèles d'attestation d'organisme bancaire ou comptable demandée au 3.1.
48. For example, (1) Schemes for the implementation of the installation of photovoltaic generation building; (2) Pictures or visual representations of the project, including images of the building before and after installation of the photovoltaic generation facility;(3) The business plan, including the duration of the purchase contract that highlights the profitability expected and detailing, at a minimum, the project's estimated revenue amounts, costs, and cash flows before and after taxes;(4) Description of the legal structure that will be used to develop the project and

– Annex 4 requires the submission of very 'technical' details related to resource composition (PolySi (/kg), Ingot wafer (/wafer), Cell (/cell), Module (/m$^2$), Front (/kg), rear (/kg), encapsulant (/kg)) of the PV module). The compulsory filing requirements include manufacturing countries and each module's energy consumption. Certain optional filing requirements include information related to $CO_2$ emissions, water consumption, toxic materials RoHS (%), manufacturing sites, and so on.[49]

The above-noted document filing requirements, particularly Annex 4, are quite bothersome. Many have asked why the bidding process employed to determine tendering prices requires the submission of unrelated detailed information. Therefore, the 2013 Simple Spec included a revised version of Annex 4. The revised Annex 4 solely requires a simple carbon footprint report. The applicant must only follow the step-by-step guide to calculate its carbon footprint in a relatively simple manner.[50]

*[b]   Review Process*

After an applicant's filing is received, energy regulators (CRE) are required to complete an application review within two months and produce a summary report. The minister will inform successful and unsuccessful applicants of bids by email. Then, the CRE will publish a list of winners on its website.[51]

*[6]   Criteria Required to Win Bids*

Under the old (2011) simple tendering scheme, the key elements required to win bids are purely based on bidding prices. Lower prices receive bids.[52] However, this price-based regime was changed. The 2013 Simple Spec integrated the concept of complex tendering that appeared in the 2011 Complex Spec. According to the new bidding regulations, the total score is 30 (i.e., 20 points awarded for price elements and 10 points awarded for simple carbon emission elements).[53] The simple carbon emission report will be based on guidelines provided by the International Energy Agency

---

to ensure the delivery of electricity. This description shall include, where appropriate, the composition of the shareholders, a list of partners involved, their roles, and the nature of their relationships to the applicant.
49. Annexe 4 Fiche déclarative relative à la constitution du laminé photovoltaïque et à la onsommation de ressources associée, 2011 Simple Spec.
50. 2013 Simple Spec, 4.2 and Annex 4.
51. 2011 Simple Spec, 2.7; 2013 Simple Spec, 2.7.
52. 2011 Simple Spec, 4 Instruction des dossiers. *See also*: Délibération du 22 mars 2012 portant avis sur le choix des offres que le ministre chargé de l'énergie envisage au terme de l'appel d'offres portant sur la réalisation et l'exploitation d'installations photovoltaïques sur bâtiment de puissance crête comprise entre 100 et 250 kW, *available online* at http://www.legifrance.gouv.fr/affichTexte.do?cidTexte=JORFTEXT000025627071.
53. 2013 Simple Spec, 5, 5.1, 5.2, and 5.3.

(IEA) for $CO_2$ emissions from fuel combustion and energy demands[54] during the production process.

However, it should be noted that limitations have been applied to the loading hours that might enjoy favorable purchasing rates.[55] Once the cap is exceeded, the rate would shrink to 5 Eurocent/kW:

- Non-sun-tracking PV: 1,500 hours (continental France); 1,800 (overseas and Corsica).
- Sun-tracking PV: 2,200 hours (continental France); 2,800 (overseas and Corsica).
- Other PV: no hourly limitations.

Yet, in the 2013 Simple Spec, only caps for Non-sun-tracking PV (i.e., 1,500 hours (continental France) and 1,800 hours (overseas and Corsica)) were mentioned.

*[7]   Procedures to Be Followed after Bids Are Won; Avoiding Delays in Construction Clauses*

Winning bids does not mean that all of the legal requirements involved in the development of PV installations have been satisfied. Successful applicants must follow the provisions of related legislations and regulations.[56]

The following additional duration requirements may apply:[57]

- Completion of the construction deadline: Applicants agree to complete installations within 18 months after they receive notifications that they won bids.
- Commercial operation duration requirements: If grid connection work conducted by transmission or distribution system operators is completed within 18 months, then deadlines will extend until 20 months after successful applicants receive notifications that they have won bids. However, if grid connection works lasts more than 18 months, then facilities must begin operations within 2 months.

The penalties for all delays are shortened purchase contract durations. This means that profits will be reduced. A one-day delay would lead to a two-day shortening of the purchase contract duration.[58] In addition, administrative penalties might be imposed.[59] Finally, any fraudulent statements provided during bidding would result in automatic contract terminations without compensation and/or reimbursement.[60]

---

54. Tableau 3: Valeurs EMj du contenu $CO_2$ du kWh électrique par pays de consommation de l'électricité provenant d'une publication de l'IAE: $CO_2$ emissions from fuel combustion, 2010.
55. 2011 Simple Spec, 3.5 Rémunération.
56. 2011 Simple Spec, 2.4; 2013 Simple Spec, 2.4.
57. 2011 Simple Spec, 3.3; 2013 Simple Spec, 3.2.
58. 2011 Simple Spec, 5.4; 2013 Simple Spec, 5.4.
59. The minister may also impose administrative and monetary sanctions for any failures by applicants to fulfill all or part of their obligations pursuant to Art. L142-31 of the Energy Code. *See also,* Simple Spec of 2013, 6.2.
60. *See,* 2013 Simple Spec, [6.2]4.2.

## [8] Results of Tendering: Five Results

Based on the competitiveness of tendering, the process can be divided into two stages.

The first stage, the learning stage, includes two phases (120 MW of 2011/8/1-2012/1/20; 30 MW: 2012/1-2012/3/31). During this stage, the results might not be quite satisfactory. Participation in the tendering scheme might not proceed as expected. For instance, the first simple tendering call attracted around half (68 MW, 345 cases) of the expected tendering capacity (120 MW). The final decisions only affected 45 MW (218 cases).[61] During the second call, the scheduled capacity was 30 MW. However, only 21 MW (109 cases) won bids.[62]

Yet, following this learning phase, applications tended to increase robustly. The scheduled capacity was also 30 MW between the third and fifth calls. However, the number of applications was 53 MW, 81 MW, and 53.8 MW, respectively.[63] However, because of caps, the authority was only able to select around 30 MW projects.[64]

## [D] 2011 and 2013 Complex Tendering Schemes

### [1] Policy Targets

The 2011 Complex Spec repeats the 2020 RE targets. However, another policy target was set between mid-2011 and mid-2012. This resulted in an annual PV target of 500

---

61. Appel d'offres–Etat des lieux: Etat des lieux sur l'appel d'offres portant sur la réalisation et l'exploitation d'installations photovoltaïques sur bâtiment de puissance crête comprise entre 100 et 250 kWc, *available online* at http://www.cre.fr/documents/appels-d-offres/appel-d-offres-portant-sur-des-installations-photovoltaiques-sur-batiment-de-puissance-crete-comprise-entre-100-et-250-kw/etat-des-lieux-de-la-premiere-periode (last retrieved 2 Aug. 2013).
62. Ministre de l'Écologie, du Développement durable et de l'Énergie, COMMUNIQUÉ DE PRESSE Nouvelles installations solaires photovoltaïques et thermodynamiques, *available online* at http://www.photovoltaique.info/IMG/pdf/2012-07-26_nles_instal-_solaires_photovoltaiques_thermodynamiques-1.pdf (last retrieved 2 Aug. 2013). *See also*, http://www.photovoltaique.info/+Des-reponses-aux-appels-d-offres+.html (last retrieved 2 Aug. 2012).
63. Délibération du 14 février 2013 portant avis sur le choix des offres que la ministre chargée de l'énergie envisage au terme de l'appel d'offres portant sur la réalisation et l'exploitation d'installations photovoltaïques sur bâtiment de puissance crête comprise entre 100 et 250 kW (3e période), *available online* at http://www.legifrance.gouv.fr/affichTexte.do?cidTexte=JORFTEXT000027150740&dateTexte=&categorieLien=id; Délibération du 14 février 2013 portant avis sur le choix des offres que la ministre chargée de l'énergie envisage au terme de l'appel d'offres portant sur la réalisation et l'exploitation d'installations photovoltaïques sur bâtiment de puissance crête comprise entre 100 et 250 kW (4e période), *available online* at http://www.legifrance.gouv.fr/affichTexte.do?cidTexte=JORFTEXT000027150745&dateTexte=&categorieLien=id; Délibération du 29 mai 2013 portant avis sur le choix des offres que la ministre chargée de l'énergie envisage au terme de l'appel d'offres portant sur la réalisation et l'exploitation d'installations photovoltaïques sur bâtiment de puissance crête comprise entre 100 et 250 kW (5e période), *available online* at http://www.legifrance.gouv.fr/affichTexte.do?cidTexte=JORFTEXT000027534670&dateTexte=&categorieLien=id.
64. For instance, the fourth result involved 143 cases (a total of 30.9 MW). The results of the fifth tender involved 139 projects (a total of 30.03 MW) (138 projects or 29.87 MW).

MW.⁶⁵ Because of this target, it was easy to determine why the first phase of simple tendering amounted to 120 MW, rather than 30 MW. Furthermore, the 180 MW target also related to the following 2014 tendering target and cap: 180 × 2.5 = 450 MW.⁶⁶ The PV development target between mid-2011 and mid-2012 are illustrated in Table 3.2.

*Table 3.2   PV Development Targets between Mid-2011 and Mid-2012*

| Not subject to tendering scheme | 100 MW/year<br>Residential PV (≤36 kW) |
|---|---|
| | 100 MW/year<br>Non-residential PV (≤100 kW) |
| Simple tendering | 120 MW/year<br>Building-related PV ( > 100 kW ~ ≤250 kW) |
| Complex tendering | 180 MW/year<br>( > 250 Kw) PV, and all ground-type PV |

*Source*: 2011 Complex Spec.

In the new 2013 Complex Spec, a new target of 400 MW (2015–2016) was set.⁶⁷ This target is equivalent to the 200 MW annual target, as well as the tendering target. This target is higher than the previous complex tendering target (180 MW).

### [2]   Tendering Targets and Caps

In comparison with simple tendering, the complicated nature of complex tendering can be determined from the diverse elements involved in the arrangement and allocation of tendering caps. The annual cap is 180 MW. This cap is higher than the cap set for simple tendering. To allocate the 450 MW cap in one tendering procedure and, taking into account the needs of diversity in PV technology,⁶⁸ this capacity has been further divided into *three big categories and seven small sub-categories*.⁶⁹ Tendering Targets and Caps in the 2011 Complex Spec are illustrated in Table 3.3.

---

65. 2011 Complex Spec, p. 1.
66. *Ibid.*, at 1.
67. 2013 Complex Spec, at 1.
68. Including thermal dynamic generation facilities. *Ibid.*, at 1.
69. 2011 Complex Spec, 3.1 Caractéristiques des installations.

Table 3.3  Tendering Targets and Caps in the 2011 Complex Spec

| Type I. Building-Related PV: 50 MW | Sub-category 1. Building-Related PV | |
|---|---|---|
| Type II. Innovative PV: 237.5 MW | 37.5 MW | Sub-category 2. Thermal dynamic facility < 37.5 MWc |
| | 50 MW | Sub-category 3. Concentrated PV < 12 MWc |
| | 100 MW | Sub-category 4. PV with sun-tracking device < 12 MWc |
| | 50 MW | Sub-category 5. PV with storage in Corsica or overseas France < 12 MWc |
| Type III. Mature PV: 162.5 MW | 125 MW | Sub-category 6. Ground-type PV (4.5 MW < P ≤40 MW) |
| | 37.5 MW | Sub-category 7. Ground-type PV (4.5 MW≤) |

Source: 2011 Complex Spec, 3.1.

The complex allocation rules that appeared in the 2011 Complex Spec were simplified in the 2013 Complex Spec. Only two big categories and five sub-categories were included in the 2013 Complex Spec: The concept of building-related technology was integrated with mature technology and the thermal dynamic facility was removed. *Tendering Targets and Caps in the 2013 Complex Spec* are illustrated in Table 3.4.

Table 3.4  Tendering Targets and Caps in the 2013 Complex Spec

| Type I. Innovative & Ground-type PV | 100 MW (Sub-category 1 concentrated PV) < 12 MWc | 1a. All concentrated PV: 20 MW |
|---|---|---|
| | | 1b Partial (at least 50%) concentrated PV: 80 MW |
| | 100 MW | Sub-category 2 PV with sun-tracking device < 12 MWc |
| Type II. Mature or Building-related PV | 60 MW | Sub-category 3. Parking lot type ground PV < 4.5 MWc |
| | 100 MW | Sub-category 4. Rooftop PV (simplified BIPV)(< 3 MWc) |
| | 40 MW | Sub-category 5. Rooftop PV (ordinary rooftop)(3–12 MW) |

Source: 2013 Complex Spec, 3.1.

*[3]   Eligibility*

The complex tendering scheme is also very complicated with respect to eligibility.

## [a] Installation Capacity

In the 2011 Complex Spec, a 250 kW–12 MW capacity range was established for building-related PV. The capacity range was set at 0 MW–4.5 MW and 4.5 MW–40 MW for ground-type PV and at 12 MW or 37.5 MW for the remaining PV types. Yet, in the 2013 Complex Spec, more delicate designs for building-related PV were established: 250 MW–3 MW, 3 MW–12 MW. Settings for ground PV remained the same (0 MW–12 MW, 0 MW–4.5 MW)

## [b] Capacity Requirements for Each Case

As noted in the 2011 Complex Spec, out of seven sub-categories, four types had 12 MW application ceilings.[70] However, three types were assigned their own individual ceilings.[71] In the 2013 Complex Spec, the 12 MW ceiling was applied to Types I-1, I-2, and II-5. Special 3 MW (II-4) and 4.5 MW (II-3) ceilings were also introduced.

## [c] Other Criteria

Many extra requirements were added for PV installations under the complex tendering scheme.

### [i] General Criteria and Requirements

The general criteria and requirements include the following factors:[72]

(1) Provision of information related to PV functions:[73] Each solar panel must securely transmit its data to a public research institute that specializes in solar energy, a cluster that specializes in solar energy, or a platform for innovation, as defined in the future investment program.
(2) ISO requirements: The manufacturer(s) of electrical equipment required for the production of materials dedicated to conversions that will provide energy, process dc and alternating currents, and/or increase voltages must provide related certificates, such as ISO 9001, ISO 14001, or equivalent certificates. This requirement is similar to the requirements that appear in the simple tendering scheme.

---

70. Sub-category 3. Concentrated PV < 12 MWc; Sub-category 4. PV with sun-tracking device < 12 MWc; Sub-category 5. PV with storage in Corsica or overseas France < 12 MWc; Sub-category 1. Building-related PV.
71. Sub-category 2. Thermal dynamic facility < 37.5 MWc; Sub-category 6. Ground-type PV (4.5 MW < P ≤ 40 MW); Sub-category 7. Ground-type PV (4.5 MW ≤).
72. 2011 Complex Spec, 3.1; 2013 Complex Spec, 3.1.
73. 2011 Complex Spec, 3.1 Caractéristiques des installations; 2013 Complex Spec, 3.1.

(3) Local government participation in the tendering process: Similar to the simple tendering scheme, local governments are permitted to participate in the tendering process.[74]
(4) Submission of reports related to Environmental Impact Assessments (EIAs) and industrial risk: For ground-type PV, both reports are mandatory. However, for certain PV types, such as rooftop PV or parking lot PV, only industrial risk reports are compulsory.[75] Detailed evaluation items are further defined in Annex 3 of the Spec. EIAs address possible impacts on landscapes and heritage, impacts on the physical environment (geology, hydrology), impacts on the natural environment (ecosystem functioning, sensitive plant and animal species), and impacts on the human environment (e.g., neighborhood nuisances during construction). Industrial risk reports focus on the prevention of intrusions, theft, malice, injuries, chemical spills, explosions, fires, damage to equipment or structures that results from malice, error handling, work accidents, potential risks for aircraft (glare), and so on. In the 2013 Complex Spec, most of these regulations are similar. With respect to ground installations, EIA reports must emphasize whether projects will be constructed on ecological restoration sites.[76]
(5) End-of-life recycling duties: Applicants agree to rehabilitate sites after operations are completed. In addition, candidates also commit to recycling used photovoltaic modules or films after operations are completed.[77]
(6) Proof of application to acquire planning permission: Applicants must provide copies of their applications to acquire planning permission to complete installations.[78]
(7) Land control requirements:[79] Applicants must secure land for PV installations during the period of operation for those installations (e.g., lease agreement or proof of ownership).
(8) Grid connection management: Applicants must file the results of detailed studies conducted by transmission operations or copies of pre-study connections.[80] In the 2013 Spec, equipment documentation has become more complex. Applicants must now provide (1) the results of preliminary studies of single network connections performed by managing networks; (2) the results of thorough pre-study network connections performed by transmission systems; and (3) Technical and Financial Proposals (TFPs) performed by network managers.[81]

---

74. Peut participer à cet appel d'offres toute personne exploitant ou désirant construire et exploiter une unité de production, sous réserve des dispositions des Articles L.2224-32 et L.2224-33 du code général des collectivités territoriales.
75. 4.3, 2011 Complex Spec.
76. 4.5, 2013 Complex Spec.
77. 3.3, 2011 Complex Spec; 3.4, 2013 Complex Spec.
78. 4.4.1, 2011 Complex Spec; 3.1, 2013 Complex Spec.
79. 42011 Complex Spec,.4.1 Maîtrise foncière et autorisation d'urbanisme; 2013 Complex Spec, 4.3.
80. 2011 Complex Spec, 4.4.2.
81. 2013 Complex Spec. 4.3.

(9) Local opinion procedures: Applicants must provide opinions of deliberations made by municipal councils of municipalities affected by projects and mayors' opinions or other local opinions to prove local acceptability.[82] However, this requirement was removed from the 2013 Spec.

(10) Proof of contributions to PV R&D: Applicants must provide information related to contributions to PV R&D in France.[83] Forms of evidence include agreements to provide related operational data to innovation platforms and related research institutes.[84] In addition, applicants must indicate potential contributions to PV R&D. These include proof of the improved performance of photovoltaic cells, improved performance of sunlight aggregating devices, improved performance of devices that track the sun's path, and so on.[85] In the 2013 Spec, this requirement was significantly expanded in section 4.6. Annex 5 addresses this issue. R&D directions focus on nine directions.[86]

(11) The provision of simple carbon evaluation reports: Applicants or authorized specialized institutes must follow the guidelines provided in Annex 5 to calculate carbon emissions. They must then submit carbon evaluation reports.[87] A similar scheme is described in the 2013 Spec.[88]

(12) Proof of applicants' project completion and implementation capacities: Applicants must demonstrate their capacities to complete the projects at stake. They must describe their previous experiences and achievements.[89] A similar requirement is included in the 2013 Spec.[90]

(13) General financial capacities and financial guarantees: Applicants must submit general information related to the financial capacities at stake. This information should include estimated investment amounts, presentations of financial packages for project equity, debts, grants, and financial

---

82. 2011 Complex Spec, 4.4.3.
83. 2011 Complex Spec, 4.5.
84. 2011 Complex Spec, 4.5.1.
85. 2011 Complex Spec, 4.5.2 Autres elements.
86. For instance, improving the performance of electrical components or component électroniques 8 a photovoltaic system; improving the performance of photovoltaic cells or modules; improving the performance of sunlight aggregating devices; improving the performance of devices that track the sun's paths; development and improvement of forecasting systems for the production of PV electricity; improving the performance of storage devices and the integration of energy PV in the network; improvement of machinery or means of production that contribute to the purification of silicon or the manufacture of solar cells or modules; improving the performance of encapsulation materials and protection cells; improvement and development of components or methods that reduce the costs of energy production.
87. 2011 Complex Spec, 4.7 Évaluation carbone simplifiée de l'installation photovoltaïque.
88. 2013 Complex Spec, 4.7 and Annex 4.
89. 2011 Complex Spec, 4.6.1.
90. 2013 Complex Spec, 4.3.

benefits, business plans; purchase contract durations that highlight expected profitability. These reports should detail, at a minimum, the estimated amounts of revenue based on project costs and cash flows before and after taxes, and so on.[91] Specifically, financial guarantees for project execution and dismantling to be performed after project completion must be secured.[92,93] A similar regime also appears in the 2013 Spec.[94] However, in the 2013 Spec, only dismantling is mentioned.[95]

Finally, in the 2013 Spec, an extra requirement was added to '2013 Complex Spec: Related Devices'. For R&D purposes, each PV panel must contain instruments that measure global irradiance incidents (both horizontally and in the modules' planes), weather conditions (temperature, wind, rain), overall production level AC (voltage, current, active and reactive power), and so on.[96]

[ii]     Criteria and Requirements for Specific PV Installations

Extra requirements were added for specific PV installations:

(1) Type I PV to be installed on buildings must include temperature sensing devices in three different locations.[97,98]
(2) Building-related PV to be installed under Sub-categories 1 and 5 must be covered by liability insurance and comply with construction codes, electricity codes, and so on.[99] However, these requirements were removed from the 2013 Spec.
(3) In the 2011 Spec, certain PV types must include energy storage devices.[100]
(4) Ground-type PV: Each solar plant must follow the provisions of Town Planning Code Laws for the modernization of agriculture and fisheries, planning laws, coastline laws, mountain laws, and so on. To avoid environmental concerns, the suggested sites include former industrial sites, old quarries, or contaminated sites.[101]

---

91. 2011 Complex Spec, 4.6.2 Structure juridique et solidité financière.
92. 2011 Complex Spec, 6.3.1.
93. 2011 Complex Spec, Annex 6 and 6.3.2 Garantie financière de démantèlement.
94. 2013 Complex Spec, 4.4, 6, Annex 6, Annex 7.
95. 2013 Complex Spec, 6.1.
96. 2013 Complex Spec, 3.1.
97. 2011 Complex Spec, 3.1 Caractéristiques des installations.
98. 2013 Complex Spec, 3.1.
99. 2011 Complex Spec, 3.1.
100. 2011 Complex Spec, 6.2 Stockage de l'énergie.
101. 2011 Complex Spec, 3.1 Caractéristiques des installations; 2013 Complex Spec, 3.1.

In addition, in the 2013 Spec, extra conditions are provided for Sub-category 1–5 installations:

(1) Sub-category 1: (1) concentration factors must exceed 400; (2) project sizes submitted by applicants and their associated affiliates must not exceed 12 MW within 500 meters of related sites.
(2) Sub-category 2: (1) The use of concentrated PV is prohibited; (2) project sizes submitted by applicants and their associated affiliates must not exceed 12 MW within 500 meters of related sites.
(3) Sub-category 3: (1) Installations must comply with the requirements for 'simple BIPV'; (2) project sizes submitted by applicants and their associated affiliates must not exceed 4.5 MW within 500 meters of related sites.
(4) Sub-category 4: (1) Installations must comply with the requirements for 'simple BIPV'; (2) project sizes submitted by applicants and their associated affiliates must not exceed 3 MW within 500 meters of related sites.
(5) Sub-category 5: (1) Parking lot PV is not allowed; (2) project sizes submitted by applicants and their associated affiliates must not exceed 12 MW within 500 meters of related sites.

### [4]    Ratios of Rooftops and Ground PV

As noted above, under the simple tendering scheme, all 300 MW is reserved for rooftop PV. However, under the complex tendering scheme, a large proportion of capacity has been allocated to ground-type PV. For instance, under the 2011 Spec, the direct (Type III, Mature PV: 162.5 MW) or indirect ground-type PV (e.g., Sub-category 3, concentrated PV < 12 MWc and Sub-category 3, parking lot type ground PV < 4.5 MWc) occupy more than half of the total capacity. Yet, only 50 MW and 1/6 has exclusively been reserved for large rooftop systems.

However, these relationships are more balanced in the 2013 Spec. Out of 400 MW, 140 MW and 35% have been reserved for Sub-category 4 (i.e., rooftop PV (simplified BIPV) (< 3 MWc)) and Sub-category 5 (i.e., rooftop PV (ordinary rooftop) (3 MW–12 MW)). Yet, more than half of the total capacity has also been allocated to ground-type PV.

### [5]    Administrative Procedures for Tendering

#### [a]    Submission of Related Documents

Documents required to prove related criteria in the complex tendering scheme have been met are much more complex than the documents required under the simple tendering scheme. According to Annex 2, the submission of six key documents is required. A similar format is required in Annex 2 of the 2013 Spec:[102]

(1) Basic application form: The detailed format is provided in Annex 1. The requirements include applicants' proof of commitment, administrative information, signatures, and so on.
(2) Project Overview: Applicants must provide ISO certificates, building permits, and so on.
(3) Applicants must provide EIAs and industrial risk reports.
(4) Applicants must provide proof of project implementation capacities.
(5) Applicants must provide proof of potential PV R&D contributions.
(6) Applicants must provide financial guarantee certificates (including construction and dismantle guarantees) issued by banks.

*[b]    Review Process*

Upon receipt of tendering applications, the CRE will open tender applications within 15 days. The CRE will quickly review the 'formality' of all applications and reject incomplete submissions. The date of opening session will not be made public.[103] The CRE will then begin to review the 'substantive' elements of all applications. The CRE must complete the selection process and transmit the results to appropriate ministers during a period that should not exceed four months from the date applications were opened. Finally, the CRE will provide candidate lists to ministers who will determine bid decisions.[104]

*[6]    Criteria to Win Bids*

According to the aforementioned simple tendering scheme, low prices are the keys to winning bids. However, according to the complex tendering scheme, the role of price has been reduced to less than half of the total evaluation consideration process (only 40%).

Under the complex tendering scheme, complexity is also reflected in its complicated evaluation scheme. A scoreboard was developed to determine differences among the various types of facilities. Differentiated reasons are based on PV types. The three main score ratings are provided in Table 3.5.[105]

---

102. 2013 Complex Spec, Annex 2.
103. 2011 Complex Spec, 2.8; 2013 Complex Spec, 2.8.
104. 2011 Complex Spec, 2.9; 2013 Complex Spec, 2.9.
105. 2011 Complex Spec, 5.1 Pondération des critères.

*Table 3.5 Ratings Used in the 2011 Complex Tendering Scheme*

| | Sub-category 1. Building-Related PV | Sub-category 3. Concentrated PV < 12 MWc Sub-category 4. PV with Sun-Tracking Device < 12 MWc Sub-category 5. PV with Storage in Corsica or Overseas France < 12 MWc Sub-category 6. Ground-Type PV (4.5 MW < Pround) | Sub-category 2. Thermal Dynamic Facility < 37.5 MWc Sub-category 7. Ground-Type PV (4.5 MW≤. |
|---|---|---|---|
| Price | 12 | 12 | 12 |
| Environmental impact, industrial risk, and carbon emission evaluation | 4 | 5 | 6 |
| Project feasibility and completion times | 7 | 5 | 6 |
| Contributions to R&D | 7 | 8 | 6 |
| Total | 30 | 30 | 30 |

*Source*: 5.1, 2011 Complex Spec.

To increase the certainty of reviewing relatively uncertain elements, the 2011 Spec provides very detailed explanations of how to review elements related to 'Environmental impact, industrial risk, and carbon emission evaluation',[106] 'Project feasibility and completion times',[107] and 'Contributions to R&D'.[108] For instance, different evaluation parameters are provided for 'Environmental factors' and 'Project feasibility and completion times' for ground-type and rooftop PV.

In the 2013 Spec, this complex evaluation table has been simplified. In addition, difficult and uncertain elements related to 'Project feasibility and completion times' have been removed. The new scoreboard is illustrated in Table 3.6.

---

106. 2011 Complex Spec, 5.3 Notation du dossier d'évaluation des impacts environnementaux et d'évaluation des risques industriels et de l'évaluation carbone simplifiée.
107. 2011 Complex Spec, 5.4 Notation de la rapidité de realization.
108. 2011 Complex Spec, 5.5 Notation de la contribution à la recherche et au développement dans le secteur solaire.

*Table 3.6   Ratings Used in the 2011 Complex Tendering Scheme*

|  | Type I. Innovative & Ground-Type PV | Type II. Mature or Building-Related PV |
|---|---|---|
| Price | 12 | 12 |
| Environmental impact, industrial risk, and carbon emission evaluation | 10 | 8 |
| Contributions to R&D | 8 | 10 |
| Total | 30 | 30 |

*Source:* 5.1, 2013 Complex Spec.

However, it should be noted that limitations have been applied to loading hours that might enjoy favorable purchasing rates.[109] Once caps are exceeded, rates would shrink to 5 Eurocent/kW:[110]

- Non-sun-tracking PV: 1,500 hours (continental France); 1,800 (overseas and Corsica).
- Sun-tracking PV: 2,200 hours (continental France); 2,800 (overseas and Corsica).
- Other types of PV: no hourly limitations.

However, in the 2013 Simple Spec, only non-sun-tracking PVs (i.e., 1,500 hours (continental France) and 1,800 hours (overseas and Corsica) are mentioned.

*[7]   Procedures to Be Followed after Bids Are Won; Avoiding Delays in Construction Clauses*

Similar to the simple tendering scheme, winning bids does not mean that all of the legal requirements involved in the development of PV installations have been satisfied. Successful applicants must follow the provisions of related legislations and regulations.[111]

In addition, extra duration requirements have been instituted.[112] Because project duration is greater under the complex tendering scheme, an extra half-year has been provided:

- Completion of construction deadlines: Applicants agree to complete their installations within 24 months after they receive notifications that they won bids.
- Commercial operations duration requirements: If grid connection work conducted by transmission or distribution system operators is completed within

---

109. 2011 Complex Spec, 4.2 Rémunération.
110. 2011 Complex Spec, 3.5 Rémunération.
111. 2011 Complex Spec, 2.4; 2013 Complex Spec, 2.4.
112. 3.2, 2011 Complex Spec; 3.2, 2013 Complex Spec.

22 months, then the deadlines will end 24 months after successful applicants receive notifications that they won bids. However, if grid connection work takes more than 22 months, then those facilities must begin operations within 2 months.

The penalty for any delays will be the shortening of purchasing contract durations. This means profits will be reduced. A one-day delay would result in a shortening of the duration by two days.[113] In addition, administrative penalties might also be imposed.[114] Finally, any fraudulent statements submitted during the bidding process would result in automatic contract termination without compensation and/or reimbursement.[115]

*[8]   Results of Tendering*

Unlike the learning curve required under the simple tendering scheme, and unlike the cold and non-ardent attitude shown to simple tendering when it was first offered, participation in the complex tendering scheme has been very enthusiastic. Even though only 450 MW capacity was originally scheduled, the scheme attracted more than five times the number of expected applications (425 cases with total capacity of 2,437 MW).[116] After excluding applications that failed to conform to application requirements, 1,891 MW and 316 cases remained for qualification through the substantive review process. The detailed results are illustrated in Table 3.7 below.

Table 3.7   Summary of Applications Received for Complex Tendering in 2012

|  | Number of Applications Received | Number of Qualified Applications | Weighted Average Price of Complete Records (€) | Total Power of Complete Records (MWp) | Desired Target Power (MWp) |
|---|---|---|---|---|---|
| Sub-category 1. Building-related PV | 80 | 50 | 210.1 | 100 | 50 |
| Sub-category 2. Thermal dynamic facility < 37.5 MWc | 2 | 1 | 349.9 | 9 | 37.5 |
| Sub-category 3. Concentrated PV < 12 MWc | 27 | 17 | 258.3 | 96.3 | 50 |

---

113. 6.3, 2011 Complex Spec.
114. 6.3, 2011 Complex Spec.
115. 6.3, 2013 Complex Spec.
116. http://www.cre.fr/documents/appels-d-offres/appel-d-offres-portant-sur-la-realisation-et-l-exploitation-d-installations-de-production-d-electricite-a-partir-de-l-energie-solaire-d-une-puissance-superieure-a-250-kwc/etat-des-lieux.

|  | Number of Applications Received | Number of Qualified Applications | Weighted Average Price of Complete Records (€) | Total Power of Complete Records (MWp) | Desired Target Power (MWp) |
|---|---|---|---|---|---|
| Sub-category 4. PV with sun-tracking device < 12 MWc | 100 | 72 | 202.9 | 562.4 | 100 |
| Sub-category 5. PV with storage in Corsica or overseas France < 12 MWc | 35 | 25 | 440.3 | 81.5 | 50 |
| Sub-category 6. Ground-type PV (4.5 MW < P ≤ 40 MW) | 79 | 67 | 188.0 | 774.1 | 125 |
| Sub-category 7. Ground-type PV (4.5 MW≤) | 102 | 78 | 197.9 | 267.3 | 37.5 |
| Total | 425 | 316 | 210.3 | 1,890.6 | 450 |

*Source*: http://www.cre.fr/documents/appels-d-offres/appel-d-offres-portant-sur-la-realisation-et-l-exploitation-d-installations-de-production-d-electricite-a-partir-de-l-energie-solaire-d-une-puissance-superieure-a-250-kwc/etat-des-lieux.

After the review process was completed, 105 projects with 520 MW won bids.[117] Why was the final decided capacity higher than the scheduled capacity? This difference was related to the failure of the simple tendering scheme's first and second phases to reach the original targets. The shortfall capacity of the simple tendering scheme's first and second phases was around 84 MW (75 MW + 9 MW). An additional 70 MW was shifted to the complex tendering scheme. The potential causes for this shortfall are quite interesting. It may have occurred because of the complexity involved in both the simple and complex tendering schemes. Because preparations must be made to address this type of red tape, it seems quite reasonable to prepare for larger projects under the complex tendering scheme, rather than to prepare for smaller projects under the simple tendering scheme.

---

117. http://www.photovoltaique.info/IMG/pdf/2012-07-26_nles_instal-_solaires_photovoltaiques_thermodynamiques-1.pdf.

## §3.03 TENDERING SCHEME IN TAIWAN

### [A] Background: The Adoption of FIT in 2009

After the Renewable Energy Act was passed in 2009, the Taiwanese government promulgated a rate schedule for FIT in 2010.[118] The PV rate schedule for 1 kW ≤ 10 kW, ≥10–500 kW, > 500 kW size installations is 11.1883, 12.9722, 11.1190 NTD, respectively.[119] This is the first rate schedule enacted by the government and related expertise in Taiwan. This schedule is primarily based on PV cost information for 2009 or early 2010. Because it fails to consider sharp PV cost reductions that occurred in 2010, the rate is considered too preferential.

### [B] The Adoption of a Combo Scheme of FIT and Tendering in 2011

This preferential rate attracted many times the number of applications (more than 800 MW) than was originally expected (75 MW).[120] Therefore, the government launched a FIT reform. On the one hand, based on the new rate Ordinance,[121] the government cut the rate dramatically. This was strongly protested by the PV industry and installers.[122] The new rates for rooftop PV of 1 kW ≤ 10 kW, ≥10–100 kW, ≥100–500 kW, > 500 kW is 10.3185, 9.1799, 8.8241, 7.9701, respectively, are equivalent to rate cuts of −29.34%, −29.23%, −31.98%, −28.32%, respectively. The new rate for all ground-type PV is 7.3297 (-34.08%).

On the other hand, Taiwan triggered the combo scheme on 17 March 2011.[123] Installations will be divided into four categories based on equipment type, installed capacity, and whether installations are self-owned or not. With the exception of Category 1 (applied capacity between 1 kW–10 kW, rooftop installation owned by the building owner) which will be subject to the unchanged FIT rate noted above, all applications from companies in the remaining three categories[124] will be subject to the tendering scheme. The government will publish calls for tendering that will contain government-set caps for each call. Then, bidders will be invited to submit their expected installed capacities and their tendering rates, which usually must be lower than the FIT rate schedule to win bids. Winners will be awarded benefits to sell

---

118. Ordinance of FIT Rate Schedule of 2010.
119. For > 500 kW installations, extra investment subsidy of 50,000 NTD is provided. Considering this factor, the FIT is around 14.6030 NTD.
120. Lin, Policy U-turn of PV Promotion Scheme, 2010-12-27, *Liberty Times*, available online at http://www.libertytimes.com.tw/2010/new/dec/27/today-e8.htm.
121. Ordinance of FIT Rate Schedule of 2011.
122. *See* Yang, The 30% Rate Cut of PV Purchasing Rate: The future is dim for PV industry, 2011-03-15, *The Economy Times*, available online at http://opencongress.ccw.org.tw/law-news/1903 (last retrieved 2 Aug. 2013).
123. PV Tendering Ordinance of Phase I of 2011.
124. (1) rooftop with an installed capacity between 1 kW–10 kW, not owned by the building owner; (2) rooftop installation with an installed capacity superior to 10 kW; (3) ground-type installation with an installed capacity superior to 1 kW that fulfills the 'Land Use Regulations'.
    *See* Art. 3 of PV Tendering Ordinance of Phase I of 2011.

electricity to Tai-Power at the same rate for 20 years. Therefore, the Taiwanese tendering scheme is actually a *FIT-rate-indexed* tendering scheme in which the rate schedule under the FIT scheme is a reference for the 'rate ceiling'. This tendering scheme seems to provide more certainty and investment security than other non-FIT related types of tendering schemes.

### [C]    Tendering Scheme

Taiwan is one of the most experienced regimes that employ tendering schemes. As of August 2013, it will have introduced $17^{125}$ tendering schemes. The tendering scheme is regulated by five administrative ordinances: the PV Tendering Ordinance of Phase I of 2011, the PV Tendering Ordinance of Phase II of 2011, the PV Tendering Ordinance of Phase III of 2011, the PV Tendering Ordinance of 2012, and the PV Tendering Ordinance (of 2013). Based on the evolution of these different ordinances, an expansive trend is apparent in the effective timeline for tendering ordinances. During the first part of 2011, three tendering ordinances were published slightly before the tendering dates. They only applied to one/several tendering schemes during a year that offered a number of tenders. This created relative uncertainty for investors. To increase certainty and to reflect the government's experience in tendering, the Ordinance of 2012 was published in advance and remained valid for the entire year. In 2013, the new tendering ordinance became effective for the remainder of that year. This means that the annual publication of new Ordinances is not required.

#### *[1]    Policy Targets*

Taiwan's renewable energy development targets were mentioned in almost every energy policy. These targets were relatively low and non-specific during the early stage. However, they recently evolved to higher and more specified levels (e.g., specific targets for PV and wind), particularly after the Fukushima accident occurred.

##### *[a]    Prior to the 2011 Fukushima Accident*

Under the Sustainable Energy Policy Framework of 2008, an increase in the ratio of low-carbon electricity in the generation mix was scheduled.[126] The scheduled target equaled more than 8% of the electricity installation mix by 2025. This target is equivalent to *6,500 MW*.

In 2009, the long-awaited Renewable Energy Act was passed. In Article 6(2) of the Act, a total policy-subsidized target was provided as a range, rather than as a fixed

---

125. One time under PV Tendering Ordinance of Phase I of 2011; six times under PV Tendering Ordinance of Phase II of 2011 and PV Tendering Ordinance of Phase III of 2011; seven times under PV Tendering Ordinance of 2012; three times under the PV Tendering Ordinance of 2013.
126. MOEA, Sustainable Energy Policy Framework of 2008, 5 Jun. 2008, *available online* at http://web3.moeaboe.gov.tw/ECW/meeting98/content/wHandMenuFile.ashx?menu_id=1431.

number: 6,500 MW–10,000 MW. In an attempt to schedule the development of different types of renewable electricity, based on Article 6(1) of the Act, the government set gradual targets to reach that goal by 2030. For instance, the scheduled targets for 2010, 2011, 2013, 2015, 2017, 2020, 2025, and 2030 were set as 35, 73, 149, 225, 300, 564, 1,000, 1,500 MW, respectively. Thus, the goal of PV is around 23% (1,500/6,500) according to this 2030 vision.[127]

In 2010, in an attempt to respond to the Copenhagen Accord, the Taiwanese government strongly encouraged renewable energy efforts by creating a new 'pack' of energy conservation and emission reduction policies. Two cross-ministerial commissions were established at the level of the Executive Yuan (equivalent to a Cabinet Office). These commissions were entitled the Commission on Energy Conservation and Emission Reduction and the Commission for the Promotion of New Energy. A new high level policy, entitled the Master Plan for Energy Conservation and Emission Reduction of 2010 was announced. Firstly, it mentions ambitious emission reduction and energy saving targets.[128] To achieve these goals, 10 parameter programs were established. With respect to renewable energy, the Parameter Programme II, 'Transforming into a Low Carbon Energy System (MOEA)', and its sub-plan, 'Project for a New Age Promotion of Renewable Energy', were created. These programs targeted the following renewable energies: PV, biomass, and wind.

On 26 March 2010, at the first meeting of the Commission for the Promotion of New Energy, the renewable energy target was made more specific. A more ambitious target was proposed for all renewable energy to be achieved by 2030. With respect to PV, a more aggressive target for 2030 was proposed: This development is predicated and expected to be 75, 145, 430, 1,250, 2,000, 2,500 MW in 2011, 2015, 2020, 2025, and 2030, respectively. This target (2.5 GW) is apparently higher than the target set for 2009 (1.5 GW).[129]

*[b]     After the 2011 Fukushima Accident*

In November 2011, the government published a 'New Energy Policy for 2011'. The policy includes four goals: secure nuclear safety, gradual reduction of nuclear power's role, creation of a low-carbon and green energy environment, and gradual achievement of the goal of a nuclear-free homeland. Naturally, renewable energy development has become one of several key policy directions. Scheduled targets are expected to increase

---

127. MOEA, the fifth FIT Review Committee of 2009, at p. 53, *available online* at http://www.moeaboe.gov.tw/Download/Policy/Renewable/meeting/files/exam05/%E7%AC%AC5%E6%AC%A1%E5%AF%A9%E5%AE%9A%E6%9C%83%E8%AD%B0%E5%A0%B1%E5%91%8A%E6%1%88%E5%8F%8A%E8%A8%8E%E8%AB%96%E6%A1%88.pdf.
128. The GHG emission rate for 2020 returns to the 2005 target; the GHG emission rate for 2025 reduces the target to the 2000 rate.
129. *See* Ma, Taiwan Renewable Energy Development Target and its Effects, *Energy Monthly*, July 2011, *available online* at http://energymonthly.tier.org.tw/Report/201107/30.pdf (last retrieved 2 Aug. 2013).

by 6,600 MW by 2025.[130] In addition, a special program entitled the 'Million PV Roofs' has been initiated and the PV target has been set at 3,100 MW by 2030. Both of these targets are higher than former targets.[131] However, it should be noted that 2011 was also the launch year for an attempt to change the original, more favorable FIT scheme to a perceived, less-preferential combo scheme. It appears that an uncoordinated and inharmonious situation exists between these ambitious targets and less favorable promotion measures.

### [2] Tendering Targets and Caps

Tendering targets and caps are summarized in Table 3.8.

Table 3.8 Tendering Targets and Caps for Tendering Schemes (2011–2013)

| Related Ordinances | Tendering Caps | Other Caps | |
|---|---|---|---|
| PV Tendering Ordinance of Phase I of 2011 | 15 MW (for this phase) | Cap for rooftop: 12 MW<br>Cap for ground-type PV: 3 MW | |
| PV Tendering Ordinance of Phase II of 2011 | 17.6 MW (for this phase) | Caps for each additional sub-phase:<br>Sub-phase 1: 5 MW (20 July 2011); Sub-phase 2:5 MW (5 August 2011); Sub-phase 3: 7.6 MW (19 August 2011) | |
| | | Cap for ground-type PV for each sub-phase: 1 MW | |
| PV Tendering Ordinance of Phase III of 2011 | 6.2 MW (for this phase) | No cap for each sub-phase:<br>Sub-phase 1 (30 September 2011); Sub-phase 2: 14 October 2011; Sub-phase 3 (21 October 2011) | |
| | | Cap for ground-type PV for each sub-phase: 1 MW | |
| PV Tendering Ordinance of 2012 | 70 MW (However, it was amended to 83 MW on 1 August 2012) | Cap for each sub-phase: 10 MW | |
| | | Sub-phase | Submission Deadline |
| | | 1 | 20 February |
| | | 2 | 19 March |
| | | 3 | 16 April |
| | | 4 | 21 May |
| | | 5 | 18 June |
| | | 6 | 16 July |

---

130. *See* MOEA, New Energy Policy of 2011, 3 Nov. 2011, *available online* at http://www.moea.gov.tw/Mns/populace/news/News.aspx?kind=1&menu_id=40&news_id=23394 (last retrieved 2 Aug. 2013).
131. *Ibid.*

| Related Ordinances | Tendering Caps | Other Caps | |
|---|---|---|---|
| | | 7 | 20 August |
| | | 8 | 17 September |
| | | 9 | 15 October |
| | | 10 | 12 November |
| | | Cap for all ground-type PV: 5 MW | |
| | | Cap for ground-type PV for each sub-phase: 2 MW | |
| PV Tendering Ordinance(of 2013) | 90 MW (However, it was amended to 83 MW on 1 August 2012.) | Cap for each sub-phase: 10 MW (However, if the total submission capacity (X) for each sub-phase is more than 10 MW, then the cap for each sub-phase can be increased to 10 MW + (X–10 MW)/2.) | |
| | | Sub-phase | Submission Deadline |
| | | 1 | 14 January |
| | | 2 | 25 February |
| | | 3 | 1 April |
| | | 4 | 29 April |
| | | 5 | 3 June |
| | | 6 | 1 July |
| | | 7 | 5 August |
| | | 8 | 2 September |
| | | 9 | 7 October |
| | | 10 | 4 November |
| | | Cap for all ground-type PV: 5 MW | |
| | | Cap for ground-type PV for each sub-phase: 2 MW | |
| | | Cap for all large rooftop PV: 15 MW. | |
| | | Cap for large rooftop PV for each sub-phase: 5 MW. | |

*Source:* Compiled by this author.

Some evolution is evident based on these developments. First, the phases of each schedule have become more certain and regulated. Second, annual caps could be increased to certain limits to encourage PV development. Third, a tendency exists to discourage ground-type PV and support rooftop PV. Finally, with respect to rooftop PV, a preference exists for smaller rooftops over larger rooftops.

*[3]  Eligibility*

*[a]  Installation Capacity*

In 2011, all ground-type PV became subject to tendering schemes. However, preference was given to rooftop PV, particularly smaller rooftop PV (between 1 kW and 10 kW, rooftop installation must be installed by building owners). The 2012 PV Tendering Ordinance loosened regulations for rooftop systems. The threshold for exemptions from tendering was set at 30 kW. Other exemptions include: (1) PV installation can be subsidized in the form of investments or demonstration subsidies provided by MOEABOE. (2) PV installations can be developed by the government or state-own companies.

In 2013, exemptions to the tendering scheme included 'all' cases that (A) involved the installation of PV on government sites and buildings owned or managed by public schools, and (B) PV installed on buildings rented to private parties that has been approved by local governments. The aggregative capacity (A + B) must not exceed 5 MW. However, for state-owned company cases, annual capacities under 5 MW are also exempted from tendering schemes.

*[b]  Capacity Requirements for Each Case*

Under three PV Tendering Ordinances passed in 2011, the capacity range for each case was set at 1 kW–2 MW. This applied to both rooftop and ground-type PV. However, in 2012, the cap for each application case (5 MW) was made more preferential for rooftop PV, than for ground-type PV (1 MW). These capacity requirements of 5 MW and 1 MW were maintained in the 2013 PV Tendering Ordinance.

*[c]  Other Criteria*

In comparison with the very complex requirements included in the simple and complex tendering schemes developed in France, the extra requirements included in Taiwan are relatively simple.

First, applicants are required to pay security deposits to participate in tendering schemes.[132] According to Article 6 of the 2013 PV Tendering Ordinance, security deposit amounts are 0.1 million, 0.2 million, 0.3 million, 0.4 million, 0.5 million and 1 million for capacity ranges of 1 kW–100 kW, 100 kW–200 kW, 200 kW–300 kW, 300 kW–400 kW, 400 kW–500 kW, and ≥500 kW, respectively.

Second, with respect to local governments' eligibility to participate in tendering schemes, local governments were allowed to participate in the bidding process under the PV Tendering Ordinance of Phase I of 2011. However, because of local

---

132. *See* e.g., Art. 6 of PV Tendering Ordinance of 2012.

governments' varying financial and subsidy situations, one local government submitted avery low price (only 75% of the FIT rate) and disturbed the tendering scheme. In this respect, since the PV Tendering Ordinance of Phase II of 2011 was passed, the PV, as well as other special subsidy schemes, including those that include compulsory 6% green elements for public buildings, and those subsidized by MOEA or MOEABOE, have been excluded from tendering schemes.[133] As noted above, exclusions or exemptions have been further expanded to other government or state-own companies,[134] and even to private parties engaged in the development of PV on governmental buildings.[135]

Finally, with respect to ground-type PV, extra requirements have been added to comply with land-related regulations.

### [4] Ratios of Rooftop and Ground-Type PV

In the PV Tendering Ordinance of Phase I of 2011, the capacity ratios between rooftop and ground-type PV ranged between 4 (12 MW) and 1 (3 MW). Since the PV Tendering Ordinance of Phase II of 2011 was passed, only installation caps for ground-type PV have been provided. However, based on this trend, a graduated higher preference for rooftop PV is apparent. For instance, under the 'three sub-phases' of the PV Tendering Ordinance of Phase II of 2011, the ratio is 5 (total cap for rooftop and ground-type PV) to 1 (1 MW for ground-type PV) or 7.6 MW (total cap for rooftop and ground-type PV) to 1 MW (ground-type PV). Under the PV Tendering Ordinance of Phase III of 2011, these figures were changed to 6.2 MW (total cap of rooftop and ground-type PV) to 1 MW (ground-type PV). In the PV Tendering Ordinance of 2012, these figures are set at 70 MW (total cap of rooftop and ground-type PV) to 5 MW (ground-type PV) (14:1) for all year caps, and 10 MW to 2 MW for each phase.

Recently, under the PV Tendering Ordinance of 2013, an even higher ratio is apparent: 18:1 [90 MW (total cap of rooftop and ground-type PV) to 5 MW (ground-type PV)] for annual caps that appear in the original schedule.[136]

### [5] Administrative Procedures for Tendering

#### [a] Submission of Related Documents

Applicants must provide related documents when they attempt to participate in the bidding process. In comparison with the complex documents required under the

---

133. Article 3 of PV Tendering Ordinance of Phase II of 2011 and PV Tendering Ordinance of Phase III of 2011.
134. Article 2, PV Tendering Ordinance of 2012.
135. Article 2, PV Tendering Ordinance of 2013.
136. The new and additional 45 MW (with a 5 MW cap for ground-type PV) was increased to the original cap in August 2013.

French simple and complex tendering schemes, the required documents in Taiwan are very simple.[137] First, applicants must submit documents that prove that future installation permissions will be provided (e.g., identity documents, site explanations, site photos, and so on).[138] Second, based on the format of Annex 3, applicants must submit proof they paid security deposits.[139] Finally, applicants must follow the format provided in Annex 1 and submit Tender Documents that indicate the 'discount rates' provided in the FIT rate schedule. These discount rates are very substantial elements that can affect the review process. In addition, with respect to ground-type PV, applicants might need to provide documents approved by land and related authorities that prove their compliance with land use regulations.

*[b]    Review Process*

After receipt of all bids (bid deadline), the government will first examine all 'formal' issues. In the event that some cases are deemed unqualified, if further information can remedy these situations, then applicants will be asked to submit related documents. All qualified cases will be subject to a second round during the substantial review process (the day bids are opened). On that day, sealed documents that indicate discount rates (i.e., 'Tender Documents') will be opened. After all bidding prices are compared, bid results will be decided and published.

It can be quite interesting to observe the evolution of different regulations related to the period that extends between the 'bid deadline', and the 'day bids are opened'. It appears that a wider gap exists between these two dates, and it can delay the entire tendering procedure. For instance, the gap only lasted one-day (28 April/29 April 2011) under the PV Tendering Ordinance of Phase I of 2011. However, it lasted more than 10 days under the PV Tendering Ordinance of Phase II of 2011 and the PV Tendering Ordinance of Phase III of 2011 (i.e., 20 July/29 July, 5 August/17 August, 19 August/31 August, 30 September/13 October, 14 October/26 October, 21 October/3 November). With respect to the 2012 and 2013 Ordinances, these gaps expanded and lasted more than 20 days. The tendering schedule for 2012 and 2013 is illustrated in Table 3.9.

*Table 3.9    PV Tendering Schedule in Ordinance of 2012*

| Sub-phase | 2012 | | 2013 | |
|---|---|---|---|---|
| | Submission Deadline | Date bids Were Opened | Submission Deadline | Date Bids Were Opened |
| 1 | 20 February | 14 March | 14 January | 6 February |
| 2 | 19 March | 11 April | 25 February | 20 March |

---

137. *See* e.g., Art. 4 of PV Tendering Ordinance of 2012.
138. *See* e.g., Art. 6 of the Ordinance on Renewable Electricity License and Registration.
139. *See* e.g., Art. 6 of PV Tendering Ordinance of 2012.

|  | 2012 | | 2013 | |
| --- | --- | --- | --- | --- |
| Sub-phase | Submission Deadline | Date bids Were Opened | Submission Deadline | Date Bids Were Opened |
| 3 | 16 April | 9 May | 1 April | 24 April |
| 4 | 21 May | 13 June | 29 April | 22 May |
| 5 | 18 June | 11 July | 3 June | 26 June |
| 6 | 16 July | 8 August | 1 July | 24 July |
| 7 | 20 August | 12 September | 5 August | 28 August |
| 8 | 17 September | 11 October | 2 September | 25 September |
| 9 | 15 October | 7 November | 7 October | 30 October |
| 10 | 12 November | 5 December | 4 November | 27 November |

*Source:* PV Tendering Ordinance of 2012; PV Tendering Ordinance of 2013.

However, based on closer observation, these gaps are critical. They avoid the unexpected preclusion of inexperienced and innocent bidders from the latter process. In reality, these efforts also make the tendering schemes friendlier. During these gap periods, if the government discovers that applicants must submit additional documents to remedy procedural flaws in their applications, it will ask the applicants to comply with these related duties. Therefore, this 'grace' period and additional measures could help normally small PV installers or household installers during the application process and reduce rejection rates caused by the filing of incomplete documentation.

### [6]  Criteria Required to Win Bids

The determination of successful bids is based on prices. In general, the lowest prices win bids.[140] However if discount rates are the same, two types of projects are prioritized: (1) smaller projects, and (2) rooftop PV is preferable to ground-type PV.[141]

Indirectly, tendering caps might also determine successful bids. Once tendering caps are reached, a *certain* (not all) extent of surplus capacity must still be qualified to win bids. For instance, under the PV Tendering Ordinance of Phase II of 2011, the cap is 5 MW. However, if the addition of a final rooftop candidate would cause caps to exceed 5 MW, then, this type of case would still be acceptable. However, if this case involves ground-type PV that exceeds 1 MW, then the maximum volume must not exceed 3% × 17.6 MW.[142]

---

140. *See* e.g., Art. 8 of PV Tendering Ordinance of 2012.
141. *See* e.g., Art. 8 of PV Tendering Ordinance of 2012.
142. Article 8 of PV Tendering Ordinance of Phase II of 2011.

## [7] Procedures to Be Followed after Bids Are Won; Avoiding Delays in Construction Clauses

The MOEABOE will issue certificates that provide proof of permission to begin construction for all successful bidders.[143] To avoid delays in the completion of construction and the initiation of operations, all successful candidates must sign contracts with Tai-Power within two months of receipt of certificates.[144] They must complete construction and receive installation registration certificates within periods specified by related Ordinances.[145] Generally, durations will end within one year after contracts are signed with Tai-Power (transmission operators and electricity buyers). Half-year extensions are possible.

As noted above, Taiwanese red tape is relatively less bothersome during the bid preparation process. However, Taiwan does require some relatively bothersome red tape during the licensing and registration process. PV installers are required to submit many documents under the Ordinance for Renewable Electricity Licenses and Registrations. For example, 10 documents must be submitted by small PV installers. They include registration formats, forms that prove the completion of construction photographs and layouts, receipts for purchases made during PV installations, and related Certificates of National Standard or ISO, and so on.[146]

## [8] Tendering Results: 18 Events

Up to this point, 18 tendering events have occurred in Taiwan. A summary of important findings are listed below:

First, with respect to *the effectiveness of the tendering schemes in achieving tendering targets,* the results of the seven tendering events conducted during the first year were unsatisfactory. In general, the number of applications was lower than the expected caps. Therefore, seven tendering events were required to reach annual targets by the end of 2011. However, after this learning stage ended, in 2012, the number of applications increased. The number of applications received during each sub-phase was higher than the scheduled caps. Targets were reached earlier than expected in August 2013 (extra capacity was provided and this event was closed in September 2013). A similar situation also occurred in 2012 by seven times bidding.

How can these situations be explained? Several reasons could be offered. First, applicants became familiar with the bidding rules and the red tape. Moreover, the government revised the PV tendering ordinances several times to make the bidding

---

143. *See,* e.g., Art. 9 of PV Tendering Ordinance of Phase III of 2011.
144. *See,* e.g., Art. 10 of PV Tendering Ordinance of 2012.
145. Article 9, the Ordinance on Renewable Electricity License and Registration.
146. Article 10 of the Ordinance on Renewable Electricity License and Registration.

procedures more convenient and predicable and to provide additional investment security. In addition, the collapse of the export market for Taiwanese PV led to the willingness of PV industries to participate when they developed local markets. In the sunny southern parts of Taiwan, local governments have provided extra incentives (e.g., loan schemes and building code preferences). Furthermore, the Million PV Roofs Programme developed under the New Energy Policy appears to have played a positive role.

Second, this article will provide a further analysis to determine whether these tendering schemes achieved their goals to address PV proliferation issues by controlling *development volume* and *costs*. With respect to development volume, this article discovered that the control volume effect was better in 2011, rather than in 2012 and 2013. In addition, controlling and cooling effect for ground-type PV was more successful than for rooftop PV.

With respect to costs, cost controls were reflected in *average discount rates*. The super-high discount rate of 25% only appears once under the PV Tendering Ordinance of Phase I of 2011. The discount rate was over 10% in only three cases,[147] even though it is usually around 5%[148] or less than 5%.[149]

## §3.04 CONCLUSION

This article provided an introduction and overview of two pioneer countries' applications of tendering schemes in 'PV' sectors that were formerly solely applied to other sectors (e.g., wind and biomass). The main contribution of this article is to identify 'Symmetrical' design elements (based on policy targets included in the Procedures to be Followed after Bids are Won: Avoiding Delays in Construction Clauses) under these tendering schemes. This article hopes these comparative elements can be further extended into PV tendering schemes in other countries, (e.g., Australia,[150] California (U.S.),[151] South Africa,[152] India,[153] and so on.

---

147. 12.74 (2012-7); 10.59 (2013-3); 16.19% (2013-4).
148. 6% (102-1); 6.95% (102-2); 5.54% (2011-3-3).
149. 2.62%, 0.31%, 2.95%, 0.31%, 3.12%, 0.31%, 3.37%, 4.40%, 4.35%, 4.72%, 2.06%, 3.20%, 2.53%, 3.09%, 4.37%.
150. *See*, e.g., Thiess-Silex Solar Consortium confirmed as major bidder to build large scale solar photovoltaic (PV) power station in Australia, 10 May 2010, *available online* at http://www.aeol.com.au/databases/news/thiess_silex_solar_sonsortium_confirmed.htm (last retrieved 2 Aug. 2013).
151. *See*, e.g., Winning bids for California's RAM come in under USD 0.089/kWh; *available online* at http://www.solarserver.com/solar-magazine/solar-news/current/2012/kw14/winning-bids-for-californias-ram-come-in-under-usd-0089kwh.html (last retrieved 2 Aug. 2013).
152. *See*, e.g., Solar Capital also successful in South African PV bid, 12 Dec. 2011, *available online* at http://www.pv-magazine.com/news/details/beitrag/solar-capital-also-successful-in-south-african-pv-bid_100005222/#axzz1uFDSFBY1 (last retrieved 2 Aug. 2013).
153. *See*, e.g., India's Karnataka invites solar bids worth 80 MW in *pv-magazine*, *available online* at http://www.pv-magazine.com/news/details/beitrag/indias-karnataka-invites-solar-bids-worth-80-mw_100003925/ (last retrieved 2 Aug. 2013).

With respect to comparisons between Taiwan and France:

(1) For *tendering targets and caps*, even if France suffers from a PV boom for several years, (based on its population), the scheduled tendering targets and capacities are several times of the amount of Taiwan's targets and caps. In addition, more French PV installations, rather than Taiwanese installations (lower than 100 kW) will participate in the FIT scheme without having to participate in the tendering process. These higher tendering caps appear to have been affected by relatively high *'policy goals'*, as well as by 'political will'.

(2) With respect to the *tendering schedule arrangements*, Taiwan originally adopted a temporary schedule it announced prior to tendering that might hamper investment security and offer limited time for applicants to address tendering red tape issues. It appears that a relatively long-term schedule (1 year in Taiwan, as opposed to 2–2.5 years in France) may increase predictability and offer more convenience to applicants.

(3) With respect to eligibility, with the exception of tendering thresholds (10/30 kW versus 100, 250 kW), in Taiwan, *eligibility* under the tendering schemes is much simpler than eligibility under even the simple French tendering scheme. With respect to installation capacity requirements, capacity requirements for each case, and other criteria, in France, these bothersome conditions have led to the development of the complex *'Administrative Procedures for Tendering'*. However, the perception that higher eligibility and more complex red tape will, in all likelihood, discourage investors might not hold true. Because similar timings are used for both simple tendering and complex tendering schemes in France, the number of applications has been much higher under the complex tendering scheme, rather than under the simple tendering scheme. Of course, this may have resulted from the 'economies of scale' effect involved in complex tendering schemes that increase investors' willingness to accept greater amounts of red tape.

(3) However, common concerns have been expressed in relation to the *installation capacity* requirements included in the eligibility requirements. Similar complaints have been expressed in France and Taiwan. These requirements (100 kW/250 kW in France and 10/30 kW in Taiwan) were criticized because they exerted favorable effects on large industries and companies and discouraged and disincentivized the development of small and medium enterprises and small PV companies.[154] To resolve this issue, a recommendation was made to raise the threshold from 100 kW to 250 kW in France.[155] In Taiwan, increasing the threshold from 10 kW to 30 kW has also partially reflected similar

---

154. *See*, e.g., ENERPLAN, Appel d'offres solaires: 'un cadre complexe pour achever un secteur moribond' Réaction d'Enerplan suite à la publication de l'appel d'offres pour des centrales solaires supérieures à 250 kW, 21 Sep. 2011, *available online* at http://www.photovoltaique.info/IMG/pdf/110920_cp_enerplan_ao_pv_sup250.pdf (last retrieved 2 Aug. 2013).
155. ENERPLAN, Appel d'offres solaires: 'un cadre complexe pour achever un secteur moribond' Réaction d'Enerplan suite à la publication de l'appel d'offres pour des centrales solaires

concerns. However, the appropriate threshold amount continues to be a very controversial question.

(4) With respect to *the ratios for rooftop and ground-type PV*: The explosive development of ground-type PV could be blamed for the PV boom. For this reason, rooftop PV has been favored under the tendering schemes in Taiwan and France. For instance, France excludes all ground-type PV under its simple tendering scheme. The major proportion of tendering capacity has been allotted to rooftop PV. However, after close examination, it appears that France's attitude toward ground-type PV, even after several years' booming effects have occurred, remains friendlier than Taiwan's attitude. For example, more than half of France's capacity has been allocated to ground-type PV (innovative PV tech and mature PV tech). Yet, it appears that ground-type PV has exerted a double-edged sword effect: On the one hand, it is a cheaper source of PV in comparison with rooftop systems. On the other hand, the loss of developmental control might cause tremendous burdens to fall on electricity consumers and affect land use. Therefore, because tendering schemes have regained developmental control of ground-type PV, it appears that its role can become even more important, even after several years' booming effects have occurred. However, Taiwan appears to be very hostile to ground-type PV. This may negatively affect the achievement of *policy targets* in 2030 and exert relatively low impacts on electricity prices.

(5) With respect to *Administrative Procedures for Tendering*, as noted above, France's procedures are much more complex than Taiwan's procedures. However, the thresholds to trigger this type of red tape are lower in Taiwan (10/30 kW/ground-type), rather than in France (100/250/ground-type).

(6) With respect to *Criteria Required to Win Bids*, Taiwan's tendering schemes and France's simple tendering scheme are similar. They are simply determined by price. However, price reduction effects in France are relatively better than the price reduction effects in Taiwan.[156] Under France's complex tendering scheme and the 2013 simple tendering of 2013, unique criteria have been determined (e.g., environmental concerns, contributions to PV R&D, project feasibility, and so on) to review cases. Because of the review process' limited public nature (black-box operations), the determinations of final decisions could become sources of 'national protectionism'. Can this novel 'innovative' approach subsidize local industries without triggering hard core 'local contents' or 'buy local' clauses and other international trade law issues?[157]

(7) With respect to the *Procedures to Be Followed after Bids are Won; Avoiding Delays in Construction Clauses*, a common concern related to the tendering

---

supérieures à 250 kW, 21 Sep. 2011, *available online* at http://www.photovoltaique.info/IMG/pdf/110920_cp_enerplan_ao_pv_sup250.pdf (last retrieved 2 Aug. 2013).
156. The discount rate in Taiwan is around 5% more at this time.
157. *See* e.g., the famous Canada Ontario FIT case. *EU v. Canada Renewable Energy Ontario's* feed-in tariff (FIT) (WTO Case #DS 426). *See also*, P.K. Rao, International Trade Policies and Climate Change Governance 19 (2012).

process is applicants' tendency to strive for bids or to 'seize vacancies first but leave them unused' effects. Therefore, the government's response usually involves it use of requirements to facilitate the completion of PV construction works. However, after closer observation, it is apparent that construction work related to PV installation in Taiwan is completed faster than it is completed in France. French tendering rules appear to provide significant amounts of leeway (18 months/24 months) for candidates to finish construction. This may tempt candidates to complete work at a slower pace. They might attempt to arbitrate between bidding offer prices and future technology cost reductions and, ultimately, earn additional profits. Considering the nature of PV, it appears that work completion requirements may be tougher in France.

Ultimately, with respect to the issue of the transformation from FIT to combo schemes, the most frequently asked question is whether this shift will impede PV development. Opponents of tendering schemes usually worry that red tape and 'caps' would cause chilling effects on PV development.[158] However, based on actual bidding results, 'mixed' proof exists. On the one hand, chilling effects did occur during the early phases of tendering schemes in Taiwan and France. (e.g., the results of Taiwanese tendering schemes in 2011, and the first and second simple tendering schemes offered in France). On the other hand, after these learning phases ended, applications began to flood in. At most, tendering is something, and something is better than nothing. It appears that the opponents of tendering schemes expect 'explosive' developments under FIT, rather than 'gradual' developments under tendering schemes. However, these opponents ignore the fact that 'explosive' development are effective only for short-term PV development. When customers, or the government, are unable to afford high electricity prices or the high costs involved in PV promotion, backfiring would cause the abolishment of the FIT scheme. Alternatively, the public's opposition might lead to the 'overnight wonder' effect for PV, such as the famous Spanish PV bubble. Thus, a combo scheme could be the best middle ground approach: It would consider the advantages and disadvantages of both schemes (i.e., FIT and tendering) and ensure 'long-term', 'sustainable' PV development, which, ultimately, is also better for overall industrial development. Additional observation and study of PV implementation through PV tendering schemes in Taiwan and France, as well as in other countries, is required to provide further evidence to determine which promotion scheme (i.e., FIT or tendering) is better. It is highly anticipated that the war will move from *RPS v. FIT to FIT v. Tendering* in the future. Will Germany and other FIT followers prove that pure-blooded FIT is better? Will France, Taiwan, and other combo scheme followers prove that hybrid and half-blooded FIT might work better?

---

158. ENERPLAN, Enerplan demande deux mesures pour relancer la filière de l'énergie solaire en France et créer des emplois, 25 janvier 2012, *available online* at http://www.photovoltaique.info/IMG/pdf/120125_cp_enerplan_deux_mesures_pour_relancer_le_solaire.pdf (last retrieved 2 Aug. 2013).

PART II The Evolution of the Existing Renewable Electricity Promotion Scheme after Fukushima Accident

CHAPTER 4
# Crossroad of FIT and RPS: What's the Next Step for China?

*Jingli Shi, Tao Ye & Jingting Yuan*

## §4.01 INTRODUCTION

Renewable Energy (RE) is the important strategic measure to reply double challenges of the increasing shortage of fossil fuel and the damage of environment. China government actively involves in reducing global climate change, keeps the energy safely, and promotes the remote area development. Through formulating RE development plan, specific RE development target was set up, and plenty of supporting policies have come on. In 2006, China started implementation of *Renewable Energy Law*,[1] and revised in 2009 based on industry development situation. New amendment was executed on 1 April 2010.[2] The key amendment is to solve the mismatch between the grid planning and construction and the RE development, and the issue of sustainable financial support for renewables. China regards Ren*ewable Energy Law* as the core, adopted a series of economic incentives, basically established a policy system including the core measures of tariff, special fund, finance subsidized loans, tax incentives, and other means.

In 2009, China government made two promises that by 2020 non-fossil fuel will account for 15% of energy consumption, and by 2020 $CO_2$ emission will reduce by 40% ~ 45% of 2005.[3] Development of RE has become one of the significant methods to realize those targets and is the inevitable choice to improve China energy safety level. Under the precondition of realizing 2020 targets, the demand of developing RE will be

---

1. PRC Presidential Ord. 33.
2. PRC Renewable Energy Law (Amendment).
3. Speech of President HU Jintao in United Nations Climate Change Conference in September 2009.

increased sharply, which proposes a new challenge for the RE in a healthy and ordered way.

China successively enacted supporting Feed-in-tariff (FITs) for biomass power generation (including refuse incineration power generation), wind power, solar PV system in 2006, 2009, and 2011. Considering the development status of different REs, the FITs for agriculture and forest residue power, refuse incineration power and solar PV modified in 2010, 2012, and 2013, respectively. Since the execution of FITs, it has become the main impetus for RE development, especially for wind energy and solar PV industry. Limited by China's electricity system and market mechanism, the choice of FIT and RPS is under controversy all the time. At present, FITs has effectively stimulated the investment of manufacturers and developers, but the current FIT system fails to balance the benefit among grid companies and other stakeholders. To eliminate the current bottleneck of China's RE development, the establishment of RPS system, which matches our national situation and policy environment, will make a significant influence on the future more large-scale RE development.

## §4.02 FEED-IN TARIFF IN CHINA

### [A] The Detailed Design of the Feed-In Tariff Scheme

China has designed FIT system for main RE generation technologies. China's FIT policy is established on three or six years tendered electricity price and comprehensive calculated cost and characteristics of each generation technologies; it mainly covers wind power, solar PV, biomass power generation, and refuse incineration power generation.

#### *[1] Technology Eligibility*

At present, China's FIT policies mainly applies to onshore wind power, solar PV, biomass power generation, and refuse incineration power generation. There is no explicit definition of renewable energy generation project scale in current regulations. Normally, China's onshore wind farms are large-scale and concentrated in Three North areas in the wilderness In the recent years, China government spurs the small-scale distributed wind farm and solar PV station in mid-east region. For the same type of technologies, no matter on-grid wind power, on-grid solar PV or on-grid biomass generation of refuse incineration power generation, the projects conform to the relative grid connection code and regulations, it can gain the legal FIT price.

*Table 4.1*

| Wind power | Enact onshore wind power FIT based on different resource areas. According to wind resource situation and project construction condition, decide to separate four wind resource areas, and enact corresponding wind power FIT. (NDRC Price. No. [2009]1906).<br><br>New onshore wind project, including coastal regions that intertidal zone is above the average high-water line in many years and islands which has fixed residents, unify to adopt the FIT of that wind resource area. The same wind farm cross the provincial boundary should adopt the same FIT, and the electricity level is referred to the higher FIT policy. (NDRC Price. No. [2009]1906).<br><br>The Tariff of Offshore wind projects in the future will be formulated by the State Council department in charge of the price separately. (NDRC Price. No. [2009]1906).<br><br>The wind project approved by the authority of provincial investment and energy need to be put on records by NDRC and NEA. (NDRC Price. No. [2009]1906). |
|---|---|
| Solar PV | Solar PV projects approved construction before 1 July 2011, up and running before 31 December 2011 without approved Tariff by NDRC, should execute the tariff of CNY 1.15/kWh (with tax). (NDRC Price No. [2011]1594).<br><br>Solar PV projects approved on and afterwards 1 July 2011, or approved construction before 1 July 2011 Yet to be put into operation before 31 December 2011, except Tibet continue to enforce the CNY 1.15/kWh tariff, in the rest provinces/area/city should execute the tariff of CNY 1/kWh. In future, NDRC will adjust the tariff in good time according to investment cost change and technology progress and other factors (NDRC Price No. [2011]1594).<br><br>Large-scale Solar PV Generation Station. According to solar resource condition and construction cost, it is separated into three solar resource areas in whole country, to design the corresponding FIT, the part that FIT is above local coal fire electricity price is paid by RE Development Fund. (NDRC Price No. [2013]1638).<br><br>Distributed solar system. according to the policy of whole electricity subsidy, the electricity price subsidy is CNY 0.42/kWh (with tax), paid by RE Development Fund, transferred by grid companies; inside, the rest electricity that the owner of distributed solar system cannot self-use will be purchased by local grid company as the price of local coal fire FIT. the self-used electricity won't be charged the different fund surcharge, and system reserve capacity fee and other relative on-grid cost. (NDRC Price No. [2013]1638). |

| Biomass | Enact FIT for agriculture and forestry biomass generation projects. New agriculture and forestry biomass generation projects that investors determined not by bidding should enact the same FIT. Through bidding to determine the investor, the tendered tariff is enacted as the original tended tariff, but the tariff should not be above the national FIT for agriculture and forestry biomass generation projects. (NDRC Price No. [2010]1579). |
|---|---|
| | The approved agriculture and forestry biomass generation project (except the bidding projects), the tariff lower than the above FIT level should be increased to the same level; if tariff higher than the national FIT level, still enact the original tariff standard. (NDRC Price No. [2010]1579). |
| | Refuse incineration generation projects for household refuse, the amount of waste is initially converted into on-grid energy, Per ton household refuse is regulated equal to 280 kWh tentatively, and enact the national FIT policy. The rest electricity generation adopts the local coal fire tariff. (NDRC Price [2012]801). |

### [2]    FIT Duration

The FITs for different renewable energy technologies were enacted step by step. For instance, the wind power FIT was enacted in 2009; FIT for agriculture and forestry biomass generation projects was enacted in 2010; FIT for solar PV was enacted in 2011, FIT for household refuse incineration was updated in 2012. All FITs are operated in short time; there is no clear duration request in the relative regulations. However, according to the pricing mechanism for power generation in China, the policy of tariff level is determined by the decision-maker according to the power plant operation time. However, there is no detailed regulation for the duration of FIT except solar PV.

Since the FIT for the wind power is enacted, Grid Company signed the purchase agreement with the wind farm annually. In some areas, the agreement will be signed for atmost five years, but there is no formal provision to show the FIT duration of wind farm. For solar PV, according to the update policy issued in August 2013, the principle FIT duration is 20 years.[4] China's FITs are determined by the relative authority according to the investment cost transformation, technology progress, and other factors.

### [3]    Tariff

### [a]    Tariff Schedule

The tariff of different renewable energy technologies are collected as below. In the early days, the renewable energy projects have the tendered tariff for wind power or FIT for

---

4. Section 4 Subs. 2, NDRC Price No. [2013]1638.

biomass power to keep the reasonable profit. For example, in 2006, the biomass power generation project could get a price subsidy for CNY 0.25/kWh, and the degression rate is 2% annually since 2010; wind power project could gain the tendered tariff and the approved tariff; the tariff of solar PV project was defined by the reasonable cost and profit, and normally one project has one tariff. After several years' experience, FIT level is enacted for different renewable energy technologies.

*Table 4.2*

|  | Enacted Time | Enacted Regulation | FIT Policy (with Tax) |
|---|---|---|---|
| Wind Power | 4 January 2006 | Renewable Energy Generation Tariff and Cost-Sharing Administration Trial Procedures | Tendered Tariff +Approved Tariff |
|  | 20 July 2009 | Notice about completing wind power on-grid electricity price | Four types of wind resource on-grid tariff: CNY 0.51 /kWh, CNY 0.54/kWh, CNY 0.58/kWh, CNY 0.61/kWh. |
| Solar PV | 4 January 2006 | Renewable Energy Generation Tariff and Cost-Sharing Administration Trial Procedures | Project tariff: reasonable cost + reasonable profit |
|  | 24 July 2011 | Notice about completing Solar PV on-grid electricity price | On-grid tariff: according to relative regulation, CNY 1.15/kWh or CNY 1/kWh |
|  | 30 August 2013 | Notice about the use of price leverage to promote the healthy development of the PV industry | Large Solar PV station: CNY 0.90/kWh, CNY 0.95/kWh, CNY 1/kWh, distributed solar PV system: the price subsidy is CNY 0.42/kWh |
| Biomass | 4 January 2006 | Renewable Energy Generation Tariff and Cost-Sharing Administration Trial Procedures | FIT: provincial FIT for coal power in 2005 plus CNY 0.25/kWh; Degression rate since 2010: Annually 2% |
|  | 18 July 2010 | Notice about completing agriculture and forestry biomass generation price | On-grid tariff: CNY 0.75/kWh |

| Enacted Time | Enacted Regulation | FIT Policy (with Tax) |
|---|---|---|
| 28 March 2012 | Notice about completing refuse incineration generation price | Per ton household refuse is regulated equal to 280 kWh tentatively, the on-grid tariff is CNY 0.65/kWh (with tax); The rest electricity generation adopts the local coal fire tariff. |

### [4] Capacity Cap

At present, there is no direct limit between China's renewable energy FIT and the scale of renewable energy generation project. The capacity of renewable energy generation project is determined by the investor and was approved by governments including National Energy Administration (NEA) or local energy administrations, according to the capacity of the project before May 2013. After May 2013, RE generation projects are approved by local governments, while NEA is in charge of the approval of total capacity plan of each provinces.

China enacted the 12th Five-Year Plan of RE Development, in which the proposed RE generation target and the later adjusted target could be regarded as the soft cap, the decided RE development layout and the approved RE project capacity by NEA basically is carried out according to it. the update target is by 2015 and 2020, wind power will reach 100 GW and 200 GW separately, by 2015 solar PV will reach 35 GW, agriculture and forest residue power 8 GW, biogas generation 2 GW, refuse incineration generation 3 GW.

The change of FIT depends on the authority charge of price to adjust by the development of market and technology progress.

### [5] Loading Hours (Resources Quality Cap)

In current renewable energy generation FIT policy, there is no cap of loading hours.

### [6] Cost Sharing and Recovery

According to Renewable Energy Law,[5] the cost of renewable generation will be shared by the electricity end-users across the whole country. In order to do this, the government of China developed policies that set up renewable surcharge on top of previous electricity rate, to cover the gap between FITs of RE power and tariff of coal power. In 2006–2011, the surcharge was charged by provincial grid companies and

---

5. REL Ch. 5 Art. 20.

managed with a separate account, earmarked for renewable generation subsidies, and could be allocated cross provincial grid enterprises. It was designed to solve the imbalanced problem of electricity subsidies brought by regional imbalanced renewable energy development. Since 2012, the RE fund was established; all RE surcharge levied in whole country should be put into the Fund account, and then the RE power enterprises should be paid.

Since 2006, renewable surcharge has been adjusted five times. In September 2013, the updated renewable surcharge is increased to 1.5 cent/kWh. So far, the National Development and Reform Commission and RE fund have released 10 rounds of 'subsidies for renewable energy tariff and quota trading scheme' with total subsidy amounting CNY 57.3 billion.

For the refuse incineration generation project, the part tariff that is higher than local coal fire on-grid electricity price should be shared in two levels. Inside, local provincial grid share CNY 0.1/kWh, the grid company can levy the additional cost to purchase renewable electricity in the sales price; the rest is paid by the national renewable energy surcharge.

*[7] Grid Connection, Usage, and Expansion Rules*

According to Renewable Energy Law,[6] grid companies are responsible for 'warranted full purchase' of renewable energy electricity. Thus, generation companies and grid companies have different explanation for the term of 'warranted full purchase'. Grid companies mainly adopt the priority scheduling to fulfill the duty of 'full purchase' of renewable energy electricity, but because of difficulties in transport and consumption, the grid is difficult to achieve truly 'full purchase' of the renewable energy electricity.

In order to encourage power companies to accept renewable energy actively, China provides appropriate subsidies for project investment and operation and maintenance costs that happen in renewable energy power generation projects connected to the grid system according to on-grid electricity amount. The subsidy standard: within 50 km, there is a subsidy of CNY 0.01/kWh; from 50 km to 100 km, CNY 0.02/kWh; above 100 km, CNY 0.03/kWh (Ministry of Finance No. [2012]102). The tariff subsidy policy for renewable energy generation on-grid project eliminates part of obstacles that the grid companies meet for the grid construction and grid renovation when building renewable energy generation projects. For example, in eastern and middle large load area where the grid network is intensive, this subsidy policy basically meets the local need. But in 'Three Northern' (including north-east, north, and north-west of China) areas, which have small load and centralized renewable energy generation projects, renewable energy electricity needs long distance transmission and consumption, the demand of subsidy for transmission line could be 10 times as much as the current subsidy standards, which means a significant inadequate.

Since 2012, China encourages the development of distributed generation system vigorously, especially the distributed solar PV system as key supporting area. In July

---

6. REL Ch. 4 Art. 14.

2013, the State Council promulgated the 'promoting the healthy development of photovoltaic industry guidance' which clearly states, distributed photovoltaic power generation connecting the user side, the enhancement of public electricity network that is caused by the on-grid of distributed system will be invested and constructed by grid enterprises.

### [B] The Results of the FIT: Implementation

Since the implementation of FIT policy, it becomes the main driving force for renewable energy industry blooming, especially for wind power and solar PV power generation. Wind power FIT policy plays a central role in the promotion of large-scale development of wind power, as well as if there was no wind power FIT policy and supporting cost-sharing policy, China's wind power development would have had difficulty to achieve high growth. China's wind power machine equipment prices declined rapidly since 2008, which is a precondition to design FIT policy of 2009; while doubling the amount of wind power for several years. By the end of 2012, on-grid wind power installed capacity reached 63 GW, generating capacity over 100 TWh, which made wind power the third power generation resource followingcoal power and hydro power.[7] The cumulative installed capacity of wind power has ranked first in the world; PV market development incentives have yielded good results; the amount of PV power plants throughout the reporting, construction increased significantly; related solar industry has been driven; solar power market turning to domestic market situation is clear, in 2012; the new Solar PV installation capacity is 4.5 GW, which was the 3rd in the world;, and power generation amounted to 3.5 TWh;[8] FIT policy on the development of agricultural and forestry waste generation also played a crucial role; other biomass, such as refuse incineration generation, biogas generation, etc., have also benefited, basically achieve the desired objectives, in 2012;, total biomass generation capacity is 7.5 GW, generation 33.7 TWh.[9]

### [C] Challenges and Solutions

FIT policy has made great achievements, but also indirectly caused some problems; the first is the cost-sharing funds problem. With the increasing size of subsidy funds, the funding gap is getting worse. With the development of renewable energy sources RE generation projects continue to expand the scale, it means the subsidy demand will be growing fast in future.

In addition, FIT policy focused more on the renewable energy supply side, but to promote networks and consumption, it is a limited role; at same time, it is also relatively weak for changes in the cost of the market responsiveness, and needs the

---

7. National Energy Administration (NEA), 2012 China Wind Power Construction Statistic Report.
8. HydroChina, 2012 China Solar Power Construction Statistic Report.
9. National Bureau of Statistics of China (NBSC), China Energy Statistic Yearbook.

government to make timely adjustments based on changes in the cost of renewable energy.

For the current solution of shortage of subsidy, China's government increased the renewable energy surcharge several times; the current 1.5 cent/kWh surcharge level can guarantee the subsidy demand before 2015, but after 2015, there probably will be some shortage. In Future, China's government probably adjusts the FIT level for some good profit renewable energy technology when the renewable energy investment cost reduces to a certain level. The indirect problem caused by FIT policy, such as the cost-sharing shortage in different areas and the reasonable electricity price level for different technologies of wind power, solar PV, and biomass generation needs to be solved by improvement of FIT policy and new policy design, such as RPS system.

## §4.03   THE NEXT STEP: THE RECENT DEVELOPMENT OF RPS IN CHINA

### [A]   Background of China's RPS Design

During 10 years' rapid development, China RE industry has obvious technical progress, the equipment manufacturing capability increased quickly, China's RE industry has entered into a whole new age of rapid scale development, which became one important component of China's electric power construction. The 12th Five-Year RE plan issued in 2012stated China's RE development targets of each RE technology by 2015; in 2013, State Council issued 'on the promotion of the healthy development of photovoltaic industry a number of opinions', further improved the photovoltaic power generation capacity target in 2015.[10] But under the current China RE incentive mechanism and power market system, realization of the above planned target still faces a series of difficulties: First, lack of the core target that can be accessed by law or policy. RE Law explicitly regulates that RE generation prior on-grid and warranted full purchase, but there is no assessment target and measurements; lack of binding assessment mechanism for power generation companies, power companies, and local governments; wind power and solar power generation accounted for the proportion of electricity consumption in China is very low in the case in some areas there have been serious abandoned wind conditions. Second, RE Law lacks specific mandatory measures. In the national energy policies, there is no priority policy to the development of renewable energy highlighting the strategic importance. Existing renewable energy development supporting policy focused on the generation side, the lack of accompanying incentive policies and measures on downstream grid enterprises, power users, and local government, and the lack of coordination among the various policies, and even conflicts, so that renewable energy-related laws and policies cannot be effectively implemented. Third, Power Technology system operational mechanism is unfit for large-scale renewable energy development. The existing power operation mechanism is for conventional power supply and the management approach is not suited to the characteristics of renewable energy power; contradiction during operation of renewable energy and

---

10. SC No. (2013)24.

conventional energy is obvious; there is a great resistance on on-grid and consumption of renewable energy.

China will design and implement RPS as major institutional innovation measures for the promotion of renewable energy development. RPS's essence is renewable energy power ratio that gives the whole society electricity consumption binding targets. For the duty-bearers of RPS, there are some arguments. Local governments, grid enterprises, and power developers (all or part) are possible to be the duty-bearers. Such can play security roles in the following aspects for renewable energy development.

First, a clear priority to develop the renewable energy in law requirements, to design clear and accessible indicators of renewable energy proportion in electricity consumption for the national and local administrative regions, to promote a precondition coordination of national energy development strategy, national and local energy planning, energy production and distribution, energy transmission operation, final energy use patterns to complete RPS.

The second is to ensure that energy-related policies must be implemented to meet the target for renewable energy development needs. RPS clearly shows that national renewable energy development strategy objectives, the local administrative regions proposed renewable energy power consumption constraints requirements, prompting government to make clear target and coordinated renewable energy policies, get rid of obstruction in renewable energy development policy and institution, requires local government must keep priority exploitation of renewable energy sources in the implementation of policies and management.

Third, to promote electricity and other energy systems turning to adapt to large-scale development and utilization of renewable energy. If the power grid enterprises as the responsibility body to complete RPS target within the coverage area, grid enterprises will be promoted to optimize the power system configuration and operation mode adjustment under the constraints acceptance of renewable energy power requirements, and actively use smart grid, energy storage technologies, demand-side intelligent power, distributed power generation and other new technologies. for the power system management, completely eradicate unjustified practices that hinder efficient use of renewable energy.

### [B] The Detailed Design of the RPS

The studies and discussions for China's RPS system design are still underway; the final introduction of the RPS policy system still needs to be more consideration. In the context of China's power system, China does not have fully competitive electricity market as Europe and the United States. In the specific design and implementation process, it is needed to design a different RPS system for the Chinese characteristics of power system and institutional mechanisms; promote the implementation of RPS system step by step, focused, orderly in China; connect renewable energy development with the existing institutional mechanisms, in order to achieve the best promoting effect.

The RPS system should be to address the renewable energy power on-grid and consumption issues as the main purpose, try to minimize conflict with existing successful policies, form a complementary relationship with existing policies and mechanisms, continue to promote renewable energy scale development. Objectives mandatory quotas in RPS policy generally combine to green certificate trading, create a renewable energy market demand, and transactions mode. although the Chinese emissions trading practice can serve as a reference. However, RPS after all has its own particularity, the specific implementation of RPS trading require further in-depth study combined with China's national conditions. It is needed to consider the actual operation situation after RPS system introduced to be accumulated.

As RPS is not formally enacted, this article mainly introduces the main reference, basic principle, development target, technology scope, and main challenge according to the domestic studies. For the detailed rate and quota-sharing design and the economy calculation among different regions, it is needed to do more measurements and assessments, which won't be described here.

### [1] Main Reference

China RPS will be established on Republic of China Renewable Energy Law Amendment, which is passed by the twelfth meeting of the Eleventh National People's Congress in 26 December 2009, clearly states that 'the whole nation implement warranted full purchase system of renewable energy generation', and requires 'according to the national RE development plan, define the proportion of RE electricity in whole generation within planning period', it means that RPS guides the RE development and utilization;

Decision about speedingup the cultivation and development of strategic emerging industries enacted by State council in October 2010 clearly states to execute RPS system, and realize the warranted full purchase regulation of new energy generation, and RPS will be regarded as major institutional innovation measures for the promotion of renewable energy development.

### [2] Basic Principles

RPS policy should follow the basic principles include:

(1) The fair distribution and rational burdens. In accordance with the level of economic development, resources, energy consumption and affordability and other factors, design fair and reasonable quotas indicators and incentive measures;
(2) Unified deployment and classified guidance. The National Energy authority guides the formulation and implementation of RPS policy. According to different regions, different companies and different renewable energy technologies development levels to design the different levels of quota system;

(3) Economic rationality and priority of effectiveness. The local government should give priority to develop local resources and consume power as much as possible locally, encourage investors priority to develop resource-rich areas of renewable energy, encourage local governments to buy off-site renewable energy generation targets to expand the scope and improve the overall acceptance capacity;

(4) Simple and workable. Policy should be designed as simple as possible, to ensure the implementation of RPS policy and reduce the cost of implementation.

RPS complexity and flexible operational mechanism is depending on national policy objectives, the electric power industry structure, management and implementation capacity, and social and political background. Under RPS design process, first of all, a clear mandatory target has to be determined, namely renewable energy development goals. Second, regulators must be clear, that is, which department is responsible for organizing, monitoring, and how to regulate. Third, the quota duty-bearers must be specified. Fourth, regulate the technical scope, namely what kind of renewable energy technology can be incorporated into green electricity quota management system.

*[3]   Development Target*

- Objectives should be lasting and gradually be increased with time.
- Five-year cycle of 'economic and social development plan,' as an important action Plan of China's reform and opening up and socialist modernization construction, defined goals, principles and overall deployment of the various stages of social and economic development. National energy planning, Power planning, and Renewable energy plan closely associated with the implementation of RPS all fit in the time period of the economic and social development planning. Therefore, the Chinese renewable energy RPS target setting should also be consistent with the time period of economic and social development plan. At the same time, to expand the influence of RPS policies, to ensure a reliable continuous growth of renewable energy generation, renewable energy development goals in RPS should also be clear in a longer period of time (such as 10 years or more) and grow in a predictable way. Reduce the cost uncertainty that quota-bearers fulfill the duty, to provide a good market environment for renewable energy development.
- *Determine the target level*: When determining the target level of development of renewable energy, at least the following factors should be fully considered: First, China's renewable energy resource use status and development potential; Second, various renewable energy generation technology development level and cost trends; Third, China economic constraints and social awareness level of renewable energy demand. In order to determine the optimal target level, a full social cost-benefit analysis is required. It is worth noting that if starting it from the stimulus purpose of development of renewable energy,

renewable energy RPS target should reach a certain level, or from international experience, it will not bring substantial public benefits.

According to the request of non-fossil fuel consumption proportion in primary energy consumption (11.4% by 2015), and the targets in the 12th Five-Year RE development plan, it is estimated that the non-hydro RE power generation will reach 310 TW, which is about 5% of the total power consumption by 2015.[11]

- *Installed capacity and generation standard*: In the RPS policy design, PRS should clearly define that renewable energy development is targeted at capacity or generation capacity. Most studies suggest that the latter is more incentive, which can ensure renewable energy generation facilities into production and its production capacity can successfully achieve the grid. Although the installed capacity is more oversight, no trace generating capacity, it cannot guarantee that renewable energy companies can produce enough power after installation facility. So even if the target is designed for capacity, there should be an appropriate capacity factor through which it can convert the quota duty into electricity.

### [4]   Technology Scope

According to Renewable Energy Law, its stipulates that 'renewable energy named in this law is wind, solar, hydro, biomass, geothermal energy, ocean energy, and other non-fossil energy sources. Hydropower applicable should be regulated by the State Council department in charge of energy and report the State Council for approval'. According to most studies about China's RPS, hydropower is not included.

### [5]   Duty-Bearers

Local governments could be one of the duty-bearers, regarding the special conditions of China. The implementation of RPS will relate to many stakeholders, and the local government can coordinate among the different stakeholders. It also reflects that the regions those consume more fossil fuel, have more responsibility to develop renewables.

Power grids are in charge of the transmission of distribution of electric power in China. Power grids, on the one hand, need to realize adding value of as state-owned enterprises, and on the other hand, provide the service to public and have a certain society responsibility.

---

11. Ren Dongming, Renewable Energy Quota System Policy Research _System Framework and Operation Machanisms, China Economy Publishing House, March 2013.

*[6]    Regulatory Agencies*

RPS regulatory authority should perform the following functions: policy development and coordination functions, to monitor consumption situation of RPS indicators, for verification on the grid companies for renewable energy grid electricity and so on. Under China's administrative organization framework, the implementation of RPS monitoring work should be charged by the national renewable energy authorities as leadership.

[C]    **Challenges and Solutions**

In 2012, NEA issued 'Notice on advice collection of RPS management approach'.[12] Collected comments from the provinces (autonomous regions and municipalities) energy authorities, the relevant state departments, the national power grid enterprises, large power generation companies, and the relevant planning and research institutes.

In future, during the implementation of RPS, there will be some important questions to be discussed, including: how to improve the regional RPS target design methodology; for the duty-bearers, defining the detailed requirement separately according to the different responsibilities; establishing the cross-regional qualified electricity generation calculation methodology and RPS statistical indicator system, guaranteeing the timely and accurate data of RPS; the feasibility of RE power generation and RPS statistic/ monitor/ assessment system in accordance with the relative state regulations, such as energy-saving, green gas emission control, and total energy consumption control; the statistical and monitoring platform that RPS need should link up with energy-saving and other energy statistical system, the pilot of RE Green certificate trading and framework exploration.

## §4.04    CONCLUSION

FIT policy in China has just started although the use of time is limited, but the FIT policy to stabilize prices and warranted purchase measures have led a rapid and effective promotion of Chinese investment in the renewable energy industry, fit for the industry early development stages. The main purpose of the implementation of the RPS is to optimize the overall development, to establish a coordination mechanisms fully taking care of the interests of all stakeholders, in order to address interface between the construction of renewable energy power and consumption market. Through the coordinated implementation of the two systems, the ultimate goal to promote renewable energy development can be realized.

Currently used classification of FIT and cost-sharing system, in nature, are price-based policy mechanism, which requires grid companies to purchase government-mandated renewable energy production of electricity. In the early development of renewable energy, it is a protective tariff to ensure that renewable electricity

---

12. NEA No. [2012]305.

sales to the higher cost of electricity from renewable energy provides a living space, eliminating uncertainty and risk that the renewable energy generation usually faced. However, FIT system for renewable energy solves the problem of electricity supply, it has no renewable energy production requirements for power generation companies, it is not necessary for renewable energy power producers to complete a mandatory amount of renewable energy goals as its obligation that makes renewable energy development goals uncertainty.

RPS is a policy mechanism which is based on number; its most notable feature is 'mandatory', main policy objective of implementation of RPS from the perspective to optimize the overall situation of development, to establish coordination mechanisms, which fully balance the interests of all stakeholders, thus solve the interface of construction of renewable energy and consumption market.

Limited by China's power institutional and policy environment, RPS policy in China at this stage can only be combined with the FIT policy together. According to domestic and foreign RPS and FIT policy practice, to carry out complementary development model research, RPS policy suitable for China's political environment coordinated with FIT policy has a very important practical significance for formulation and implementation of renewable energy policy in China.

Under RPS design process, it is needed to take full account of complex relationship among government policy makers, grid companies,and power generation enterprises. In the promotion process, it should be executed around RPS target decomposition and the supervision and evaluation work, as well as the establishment of coordination mechanisms to balance interests in power system. Meanwhile, it should resolve the overlapping functions among the government departments, to ensure smooth flow of information in the oversight process, which can improve the new system continuously.

To sum up, there are two advanced options that China can adopt in the current stage, RPS and FIT. It is not necessary to choose one system, can use combining institutional model of two means, which provide possibility for complementary model design.

CHAPTER 5
# Renewable Energy Development in the Philippines: Legal Measures, Implementation, Challenges, and Solutions

*Manuel Peter S. Solis*

## §5.01 INTRODUCTION

The Philippines is an archipelagic country comprising 7,107 islands scattered across the Pacific seascape of tropical Southeast Asia. It has a total land area of approximately 300,000 square kilometers and a coastline that is 36,289 kilometers long. The main island groups are Luzon, Visayas, and Mindanao. With close to 97 million people[1] and an average annual population growth rate of 1.9%,[2] the Philippines is one of the most populous countries in the region. In recent times, the country has been enjoying an economic renaissance with real Gross Domestic Product (GDP) predicted to rise at least 5% per annum from 2011 to 2030.[3] In 2012, the full-year GDP expanded by 6.6%.[4] However, the poverty headcount ratio at the national poverty level remains significant at 26.5%.[5]

With the upward trajectory of the Philippine economy in the next two decades, a concomitant increase in energy consumption is expected. Numerous studies indicate

---

1. World Bank, *Data: Philippines*, http://data.worldbank.org/country/philippines#cp_wdi (accessed 4 Sep. 2013).
2. National Statistics Office of the Philippines, *National Quickstat As of January 2013*, http://www.census.gov.ph/sites/default/files/attachments/ird/quickstat/January2013.pdf (accessed 4 Sep. 2013).
3. The Institute of Energy Economics Japan, ASEAN Centre for Energy and National ESSPA Project Teams, *The 3rd ASEAN Energy Outlook* (2011), 61, http://aseanenergy.org/media/filemanager/2012/06/14/t/3/t3aeo-complete-outlook.pdf (accessed 4 Sep. 2013).
4. National Statistical Coordination Board, *National Accounts of the Philippines – Press Release*, http://www.nscb.gov.ph/sna/2012/4th2012/2012qpr4.asp (accessed 4 Sep. 2013).
5. World Bank, *supra* n. 1.

that there is a strong correlation between economic growth and rise in energy consumption, especially if the development momentum is to be sustained.[6] Pertinently, electricity consumption is seen to grow fastest, which is predicated on the growing reliance by the industrial, commercial, and household sectors on electricity while moving away from traditional fuels.[7] In the next 20 years, attention is also drawn to the impact of total household electrification, which will intensify electricity demand.[8] Unfortunately, the cost of electricity in the Philippines is the second highest in Asia, which is mainly attributed to the high intrinsic cost of supply.[9]

To achieve its policy goals towards improving energy security, promoting investments in the energy sector, enhancing energy access, and promoting a low carbon future,[10] the Philippines has enacted the Renewable Energy Act of 2008 (REA). In particular, it declares as a matter of State policy to adopt sustainable energy development strategies to wean away the country from its dependence on fossil fuels by accelerating the exploration and development renewable energy sources.[11] Five years after the passage of the REA, however, several implementation gaps, such as the delayed and fragmented issuance of the Feed-in-Tariff (FIT) and Renewable Portfolio Standards (RPS) rules, are observed that stymie the development of the renewable electricity sector in the Philippines. Along this line, this paper aims to identify the implementation weaknesses mainly associated with the FIT and RPS since the enactment of the REA. It then concludes that only the full, timely, coherent, and consistent implementation of the FIT and RPS Schemes, among others, will the Philippines truly reap the benefits that renewable electricity offers in attaining the policy aspirations and goals enshrined in the REA.

### [A] The Energy Situation

According to Japan's Institute of Energy Economics and the Association of Southeast Asian Nations Centre for Energy, the country's primary energy consumption growth rate rose to 2.3% from 1990 to 2007.[12] Notably, the Philippine economy is mainly spurred by services (trade, transport, real estate, finance, communications, and private/government services) and industry (manufacturing, construction, mining,

---

6. The Institute of Energy Economics Japan, ASEAN Centre for Energy and National ESSPA Project Teams, *supra* n. 3.
7. *Ibid.*, 63.
8. *Ibid.*
9. International Energy Consultants, *Regional Comparison of Retail Electricity Tariffs Executive Summary* (2012), http://www.meralco.com.ph/pdf/newsandupdates/2012/NW04812a_link.pdf, 14, (accessed 5 Sep. 2013).
10. Department of Energy, *Philippine Energy Plan 2012-2030*, http://www.doe.gov.ph/doe_files/pdf/01_Energy_Situationer/2012-2030-PEP.pdf (accessed 5 Sep. 2013).
11. Department of Energy, *Issuances*, Republic Act No. 9513, s. 2(a), http://www.doe.gov.ph/issuances/republic-act/627-ra-9513 (accessed 5 Sep. 2013).
12. The Institute of Energy Economics Japan, ASEAN Centre for Energy and National ESSPA Project Teams, *supra* n. 3, 67.

electricity, and water), which are both energy-intensive sectors.[13] Under a business-as-usual scenario, the trajectory of total final energy consumption is predicted to go upwards to 4.4% until 2030.[14] However, oil and coal are still seen as major primary energy sources of supply in meeting energy consumption demand until 2030.[15] This, in turn, makes the electricity generating sector the highest contributor to greenhouse gas emission in the country representing close to 40% of the total.[16] Considering that oil and coal are mainly sourced externally, the Philippines will remain highly reliant on fossil fuel imports to cover its present and future needs in the absence of innovative policy reforms to the contrary.

To achieve energy security, the Philippines recognized the need to tap indigenous energy sources for adequate and reliable electricity supply, including the acceleration of rural electrification.[17] This is a critical strategy to inclusive growth as recurring power shortages; especially in Mindanao imperil development in the Southern part of the country.[18] So far, the percentage share of renewable energy in electricity generation in the country is encouraging with geothermal and hydropower contributing more than 17% and 14% to the total primary energy mix, respectively.[19] In sum, renewable energy supplied about 41% to the total primary energy mix in 2011.[20] However, the Asian Development Bank predicts that the share of renewables in the country will fall to 14% in 2035, at the same time that the Philippines' proven indigenous gas and coal reserves will be depleted.[21]

### [B] Renewable Energy Sources and the National Renewable Energy Program

The Philippines has vast potential to harness power from a number of renewable energy sources such as solar, wind, biomass, micro/mini hydropower, and ocean. As enunciated by the Department of Energy (DoE) under the National Renewable Energy Program (NREP), the Philippines seeks to triple its existing renewable energy portfolio

---

13. *Ibid.*
14. *Ibid.*
15. *Ibid.*, 62.
16. Greenpeace, *Green is gold: How renewable energy can save us money and generate jobs* (2013), http://www.greenpeace.org/seasia/ph/PageFiles/481216/Green-is-Gold-How-Renewable-Energy-can-save-us-money-and-generate-jobs-03.pdf, 17, (accessed 7 Sep. 2013).
17. National Economic Development Authority, *Socioeconomic Report: The First Two Years of the Aquino Administration 2010-2012* (2012), http://www.neda.gov.ph/econreports_dbs/SER/SER2010-2012new.pdf, 96, (accessed 5 Sep. 2013).
18. World Bank, *Philippine Economic Update: Accelerating Reforms to Sustain Growth* (2012), http://www.worldbank.org/content/dam/Worldbank/document/ph_Philippine_Economic_Update_Dec2012.pdf, 37, (accessed 6 Sep. 2013).
19. Samantha Olz and Milou Beerepoot, *Deploying Renewables in Southeast Asia: Trends and Potentials* (2010), http://www.oecd-ilibrary.org/docserver/download/5kmd4xs1jtmr.pdf?expires=1378533700&id=id&accname=guest&checksum=AF29F6DA6839978E0426ACBED5ECB B65, 26, (accessed 7 Sep. 2013).
20. Department of Energy, *Philippine Energy Plan 2012-2030*, http://www.doe.gov.ph/doe_files/pdf/01_Energy_Situationer/2012-2030-PEP.pdf (accessed 5 Sep. 2013).
21. Asian Development Bank, *Asian Development Outlook 2013: Asia's Energy Challenge* (2013), http://www.adb.org/sites/default/files/pub/2013/ado2013.pdf, (accessed 5 Sep. 2013).

by targeting additional capacity of 9,931.3 MW by 2030.[22] To appreciate the potential contribution of renewable energy to the total primary energy mix and in enhancing energy security in the country, the above identified renewable energy sources are briefly discussed as follows.

*[1]   Solar*

There is an enormous opportunity to utilize solar technologies in the Philippines. In the next two decades, the DoE plans to install 350 MW of additional capacity by 2030 from solar per the indicative interim target set in the NREP.[23] Interestingly, there is a degree of confidence in using solar power systems in the country for several reasons. Aside from being globally tried, tested, and mature technology, solar power systems have become the 'third most important RE in terms of global installed capacity.'[24] Also, solar power systems bring a slew of environmental returns with zero emissions or damage to the environment. With an annual potential average of 5.1 kWh/m$^2$/day,[25] solar power can also be harnessed in the country at any given scale (small, medium, or large), whether centralized or distributed, with relatively short completion time, and in an unobtrusive manner.[26]

The deployment of solar power technologies in off-grid *barangays, sitios,* and households is of particular interest that can spur livelihood and economic activities in the countryside. Many rural electrification projects undertaking 'last mile' electrification such as the DoE's Rural Power Project found solar power systems to be ideal in providing basic electricity services such as lighting and powering small appliances (mobile phone, television, radio, and computers) in off-grid and remote rural areas. However, there is a need to strengthen sustainability mechanisms such as community ownership and operatorship of the solar power systems in isolated areas with strong technical and financial support from the government.[27]

---

22. Secretary Carlos Jericho L. Petilla, *Presentation of the Philippine Energy Plan 2012-2030*, http://www.doe.gov.ph/doe_files/pdf/Researchers_Downloable_Files/EnergyPresentation/PEP_2012-2030_Presentation_(Sec_Petilla).pdf (accessed 5 Sep. 2013).
23. *Ibid.*
24. Deutsche Gesselschaft fur Internationale and Federal Ministry of Economics and Technology, *It's More Sun in the Philippines: Facts and Figures on Solar Energy in the Philippines* (2012), 4.
25. Department of Energy, *Renewable Energy*, http://www.doe.gov.ph/ER/BioOSW.htm (accessed 5 Sep. 2013).
26. Deutsche Gesselschaft fur Internationale and Federal Ministry of Economics and Technology, *supra* n. 24, 5.
27. World Bank and The Energy and Mining Sector Board, *Designing sustainable off-grid rural electrification projects: Principles and Practices* (2008), http://siteresources.worldbank.org/EXTENERGY2/Resources/OffgridGuidelines.pdf, 14, (accessed 5 Sep. 2013).

## [2] Biomass

As a predominantly agricultural country, the Philippines has vast potential (235 million barrels of fuel oil equivalent)[28] to generate power from agricultural wastes or biomass coming from sugar cane or bagasse, coconut, abaca, and rice husks, including landfill waste. Per the NREP, the DoE intends to expand its biomass portfolio by adding 276.7 MW to existing capacity by 2015, albeit there is no indicated expansion from 2016 until 2030.[29] According to the Industry Studies Department of the Board of Investments, biomass technology is a mature, adaptable and modular technology that can be used for distributed generation in all areas in the country.[30] With the abundance of cheap feedstock from biomass resources, the Board of Investments notes that electricity generated from biomass is cheaper compared to more capital intensive geothermal and hydro power plants.[31] It is also boasted that biomass potentially reduces 1.78 million tons of carbon dioxide emission, generates employment and improves rural economies.[32] As of 2009, biomass supplied more than 13% to the country's total primary energy mix.[33] Locations found ideal for biomass supply and plant facilities are in Central Luzon for rice hull; the Visayas for bagasse; and coconut residues in Southern Luzon and parts of the Visayas and Mindanao islands.

## [3] Hydropower

With an average annual rainfall of 2,360 millimeters and a tropical weather, the Philippines is well-positioned to explore and develop micro and mini hydropower in its 421 principal river basins that drain into 1,400 square kilometers of watersheds.[34] However, it is observed that the long dry season of six months makes hydropower a questionable year round and intermittent source of power generation.[35] Since run-of-river hydro facilities are typically situated in far-flung and less developed areas of the country, they have the potential to spur economic and livelihood activities in local communities.[36] In 2009, it was reported that micro and mini hydropower supplied 16% to the total primary energy mix in the country.[37] As indicated in the NREP, the DoE seeks to ramp up hydropower capacity with the addition of 5,394.1 MW by 2030.[38] This is likely to be exceeded considering that the country's hydropower development

---

28. Fernando Roxas and Andrea Santiago, *A New Systems Paradigm for the Rural Electrification Program, Philippines*, http://www.worldenergy.org/documents/congresspapers/159.pdf, 13, (accessed 5 Sep. 2013).
29. Secretary Carlos Jericho L. Petilla, *supra* n. 22.
30. Board of Investment, Biomass at http://www.boi.gov.ph/index.php/en/doing-business/industry-profiles/renewable-energy/biomass (accessed 23 Jan. 2014).
31. *Ibid.*
32. *Ibid.*
33. Greenpeace, *supra* n. 16, 32.
34. *Ibid.*, 28.
35. *Ibid.*
36. *Ibid.*
37. *Ibid.*
38. Secretary Carlos Jericho L. Petilla, *supra* n. 22.

potential has been estimated to be in the vicinity of 10,500 MW.[39] So far, there are 70 hydropower projects that have the potential to add 2,604 MW to the total primary energy mix.[40]

### [4] Wind

Being on the fringes of the Asia-Pacific monsoon belt, the Philippines has great potential to harness power from wind energy resources, particularly in the northern parts of Luzon.[41] Previous studies reveal that the country's wind density reaches a mean average of about 31 watts per square meter and over 10,000 square kilometers of windy land areas with a good-to-excellent wind resource potential.[42] This theoretically translates to more than 70,000 MW of potential installed capacity.[43] Similar to solar and biomass, wind energy is scalable and can be centralized or distributed. In 2005, the Philippines had its first wind farm – a 33 MW grid-connected facility located in Northern Luzon. Since then, six additional wind farm projects are in the pipeline with potential to add another 275 MW to the total primary energy mix.[44] Under the NREP, the DoE seeks to increase wind capacity by adding 2,345 MW by 2030.[45] So far, wind power resources are currently harnessed in the country to augment power supply to the national grid system rather than for off-grid electrification purposes.

### [5] Ocean

The Philippines has an estimated ocean resource area of 1,000 square kilometers and potential theoretical capacity of close to 170,000 MW.[46] While there is still no ocean renewable energy facility in the country, there is potential for ocean energy to contribute substantially to the total primary energy mix. Due to its vast availability and high predictability, it is seen as a viable renewable energy resource that can be harnessed through low visual impact and no $CO_2$ emission technologies such as ocean thermal energy conversion.[47] It is noted that the few fully operational commercial and tidal generating plants in the world are grid-connected.[48] At this point, the DoE has indicated in the NREP its intention to put in place an additional capacity of 70.5 MW from ocean by 2030.[49] So far, the DoE has initially identified ocean energy potential sites in the Hinatuan Passage, Camarines, Northeastern Samar, Surigao, Batan Island,

---

39. *Ibid.*
40. *Ibid.*
41. Greenpeace, *supra* n. 16, 26.
42. Department of Energy, *supra* n. 25.
43. *Ibid.*
44. Greenpeace, *supra* n. 16, 26.
45. Secretary Carlos Jericho L. Petilla, *supra* n. 22.
46. Department of Energy, *supra* n. 25.
47. Greenpeace, *supra* n. 16, 34.
48. *Ibid.*
49. Secretary Carlos Jericho L. Petilla, *supra* n. 23.

Catanduanes, Tacloban, San Bernardino Strait, Babuyan Island, Ilocos Norte, Siargao Island, and Davao Oriental.[50]

## §5.02 THE RENEWABLE ENERGY ACT OF 2008

In 2008, the REA was enacted to provide a national framework for the promotion, development, utilization, and commercialization of renewable energy sources in the country. In essence, it mandates the adoption of sustainable energy development strategies as a matter of State policy to lessen the country's dependence on fossil fuels by accelerating the exploration and development of renewable energy sources.[51] Also, the law sets out the institutional arrangement and the fiscal and non-fiscal incentives available for on-grid and off-grid renewable energy development, including various schemes and mechanisms to support renewable energy development, utilization and commercialization.

Specifically, the DoE has been designated as the lead government agency for the implementation of the provisions of the REA. To support the DoE, the Renewable Energy Management Bureau (REMB) is created as a staff and support bureau of the DoE to implement policies, plans and programs to accelerate the development, utilization and commercialization of renewable energy resources and technologies.[52] Also, REMB is empowered to develop and maintain a national information database on renewable energy sources, undertake technical research, conduct socio-economic and environmental impact studies, and ensure compliance with rules, regulations, guidelines and standards on renewable energy resources development and utilization.[53]

Aside from the REMB, the National Renewable Energy Board (NREB) is established to report directly under the DoE Secretary or Undersecretary.[54] The NREB consists of multi-sector representatives from different government line agencies, government-owned or controlled corporations and financial institutions, renewable energy developers, private distribution utilities, electric cooperatives, electric suppliers and non-governmental organizations.[55] Among its functions, the NREB is tasked to evaluate and recommend to the DoE the Renewable Portfolio Standard (RPS)[56] and the minimum renewable energy generation capacities in off-grid areas.[57] It is also empowered to recommend specific actions, monitor and review the implementation of the NREP, which seeks to attain consistency and preclude functional overlaps among the different government agencies involved in renewable energy development.[58] Moreover, the NREB is mandated to oversee and monitor the Renewable Energy Trust Fund,

---

50. Department of Energy, *supra* n. 25.
51. Department of Energy, *supra* n. 11, s. 2.
52. *Ibid.*, s. 32.
53. *Ibid.*
54. *Ibid.*, s. 27.
55. *Ibid.*
56. *Ibid.*, s. 4(ss).
57. *Ibid.*
58. *Ibid.*, s. 27.

albeit administered by the DoE, to fund resource and market assessment, research, development, demonstration, and promotion of renewable energy systems.[59]

### [A] RPS

The RPS obligates electricity industry participants such as generators, distribution utilities, or suppliers to source or produce a minimum percentage of their power requirements from eligible renewable energy resources on a sector and per grid basis.[60] This is intended to diversify the supply of energy, while at the same time reducing greenhouse gas emissions in the country.[61] To facilitate compliance with the RPS, the DoE is authorized to create a Renewable Energy Market (REM) and to supervise the establishment of a Renewable Energy Registrar (RER) through the Philippine Electricity Market Corporation (PEMC). The RER can issue Renewable Energy Certificates (REC) as proof of compliance with the RPS, which can then be traded in the REM.[62] Renewable energy certificates can also form part of an international trading emission and compliance scheme. However, the issuance of the RPS Rules by the DoE is still being awaited as of this writing.[63]

### [B] FIT

The REA mandates the implementation of a FIT system for electricity from emerging renewable energy sources such as wind, solar, ocean, run-of-river hydro, and biomass. The FIT obligates electricity power industry participants to source electricity from emerging renewable energy sources at a guaranteed fixed price for a period of not less than 12 years.[64] In addition, it provides for priority connection to the grid for electricity generated from emerging renewable energy sources as well as priority purchase and transmission of such electricity by grid system operators.[65] This means that generation for own use is excluded. Disturbingly, the process of issuing and promulgating the FIT Rules had been quite tedious taking two years from effectivity of the REA to complete, instead of within the one year window provided to the Energy Regulatory Commission (ERC) – an independent quasi-judicial regulatory body with rate fixing powers – under the law.[66] It almost took another two years to announce the first round of tariffs in July 2012, including the target installation per emerging renewable energy technology. And

---

59. *Ibid.*
60. Department of Energy, *Issuances*, DC 2009-05-0008-Rules and Regulations Implementing Republic Act No. 9513, s. 4, http://www.doe.gov.ph/doe_files/pdf/issuances/DC/DC2009-05-0008.pdf. (accessed 5 Sep. 2013).
61. *Ibid.*
62. *Ibid.*, s. 8.
63. Myrna Velasco, Renewable energy's twisted policy track (Part One), *Manila Bulletin*, 5 Apr. 2013 at http://www.mb.com.ph/article.php?aid=6289&sid=2&subid=77#.UV6-Gr9RroA, (Accessed 6 Apr. 2013).
64. Department of Energy, *supra* n. 60, s. 5.
65. *Ibid.*, s. 7.
66. *Ibid.*

lastly, the issued FIT rules only cover on-grid renewable energy systems[67] with separate implementation of the incentive mechanisms for off-grid areas.[68] There are still no FIT rules for off-grid renewable energy systems up to the present.

### [C] Green Energy Option

The DoE is mandated to establish a Green Energy Option program that allows end-users to choose renewable energy as their source of power.[69] Subject to the determination of the DoE, end-users may directly contract from renewable energy facilities their energy requirements through the relevant distribution utilities.[70] There are two simultaneous issuances that are necessary to the implementation of the Green Energy Option program. While the DoE 'shall, upon consultation with the NREB, promulgate the appropriate implementing rules and regulations which are necessary, incidental or convenient to achieve the objectives of the Green Energy Option program,'[71] the ERC – an independent regulatory body – 'shall issue the necessary regulatory framework to effect and achieve the objectives of the Green Energy Option program'[72] within six months from effectivity of REA's Implementing Rules and Regulations (IRR), i.e., from June 2009. An end-user who chooses to enroll in the Green Energy Option program must be informed by way of monthly electricity bills on how much is consumed from, and the generation charge provided by, renewable energy facilities.[73] Both issuances from the DoE and ERC remain pending.

### [D] Net-Metering

Net-metering is adopted as a consumer-based renewable energy incentive scheme wherein distribution end-users generate electricity from an eligible on-site renewable energy facility that is delivered to the local distribution grid.[74] The electricity generated can then be used by distribution end-users to offset electricity consumed from the distribution utility, or gain credit in case of electricity delivered to the grid from the on-site renewable energy facility exceeds what is consumed therefrom.[75] The distribution utility is required to enter into a net-metering agreement upon request of a distribution end-user wishing to install an on-site renewable energy facility, subject to the distribution utility's technical standards, including economic considerations, for

---

67. Energy Regulatory Commission, Resolution No. 16, Series of 2010, Rule 2.1, http://www.erc.gov.ph/admin/UploadFiles/Documents/ResolutionNo16_s_2010_final_FITrules.pdf (accessed 5 Sep. 2013).
68. Ibid.
69. Department of Energy, *supra* n. 11.
70. Ibid., s. 9.
71. Department of Energy, *supra* n. 60, s. 6.
72. Ibid.
73. Ibid.
74. Department of Energy, *supra* n. 60, s. 7.
75. Ibid.

the renewable energy facility.[76] To make the scheme more attractive to the distribution utility, it will be entitled to any renewable energy certificate issued under the arrangement, which in turn can be counted towards its compliance with the RPS.[77] Accordingly, the ERC is mandated to establish the net-metering interconnection standards, pricing methodology, and other commercial arrangements necessary to ensure the success of such a program within one year from effectivity of REA.[78] On 27 May 2013, the ERC issued the net-metering rules or guidelines albeit delayed by more than four years since the REA took effect in 2009.

[E]     Fiscal Incentives

In general, the fiscal incentives available to renewable energy developers include income tax holiday for the first seven years of commercial operation of a renewable energy facility, duty-free importation of renewable energy machinery, equipment and material, special realty tax and preferential corporate tax rates, zero value-added tax rate for sale of fuel or power generated from renewable energy sources, and tax exemption of carbon credits, among others.[79] Also, fiscal incentives for renewable energy commercialization are extended to all manufacturers, fabricators and suppliers of locally-produced renewable energy equipment and components.[80] To avail of the fiscal incentives, renewable energy developers and local manufacturers, fabricators, and suppliers must register with the DoE through the REMB, which correspondingly issues a certification. They also need to comply with the requirements, if any, imposed by other relevant government agencies charged to administer the fiscal incentives under the REA.[81]

## §5.03    THE FEED-IN-TARIFF SCHEME

[A]     Coverage

The FIT Rules cover renewable energy power plants utilizing wind, solar, ocean, run-of-river hydro, biomass, and a combination or hybrid system of the foregoing energy sources.[82] Also, the existing FIT Rules apply only to on-grid areas with separate issuance of the incentive mechanism for off-grid areas.[83] It will be noted that the FIT is initially established as a fixed tariff – not as a premium – albeit the ERC can later adopt a premium-based FIT when deemed appropriate.[84] To be eligible, a renewable energy

---

76. Ibid.
77. Ibid.
78. Department of Energy, *supra* n. 11, s.10.
79. Ibid., s. 15.
80. Ibid., s. 21.
81. Ibid., s. 25.
82. Energy Regulatory Commission, Feed-in Tariff Rules, Resolution No. 16, Series of 2010, Rule 3.
83. Ibid., Rule 2.1.
84. Ibid., Rule 2.3.

power plant has to secure a Certificate of Compliance from the ERC to avail the appropriate FIT.[85]

### [B] Duration

An eligible renewable energy power plant, which is an electricity generator from solar, biomass, wind, and hydro sources, is currently entitled to receive the applicable FIT for a period of 20 years. Notably, any renewable energy power plant that commenced commercial operation after the effectivity of the REA and were not bound under any contract to supply the energy generated to any distribution utility or consumer may still avail the FIT for 20 years less the number of years it has been in operation.[86] Also, the ERC is empowered to determine the appropriate duration after implementation of the initial FIT, which in no instance must be lower than 12 years as provided under the REA.[87] After the lapse of the FIT duration, renewable energy power plants that continue to operate are expected to base their tariffs on prevailing market prices or according to agreed prices with corresponding off-takers.[88]

### [C] Installation Target, FIT Rate, and Degression Rate

In July 2012, the ERC approved the FIT rates for four renewable energy technology systems: wind, biomass, solar, and hydro, including the degression rate for each technology.[89] The 'degression rate' refers to the rate of reduction of the FIT as the renewable energy technology matures and the cost associated with it declines over time.[90] The approved FIT rates also indicated the installation target for each technology type pursuant to the DoE's approved installation targets. It will be noted that the installation targets are not intended to be limits or caps.[91] If the initial installation target per technology is reached, it triggers the ERC to review and re-adjust the FIT.[92] However, any re-adjustment to the FIT in such instance shall only apply to new renewable energy projects with the prevailing FIT still applied to those already existing and eligible at the time of the approval of the new FIT.[93] While there is an installation target of 10 MW for ocean, no corresponding FIT rate has been issued by the ERC for power generated from this renewable energy source.

The following Table 5.1 summarizes the technology type, installation target, FIT rate, and degression rate:

---

85. *Ibid.*, Rule 1.4.
86. *Ibid.*
87. *Ibid.*, Rule 4.
88. *Ibid.*
89. Energy Regulatory Commission, Resolution No. 10, Series of 2012, Approving the Feed-in Tariff Rates.
90. Energy Regulatory Commission, *supra* n. 82, Rule 1.3.
91. Pete Maniego Jr., *Status of the RE Mechanisms: Sharing Experiences on RE Promotion* (27 Feb. 2012), http://eeas.europa.eu/delegations/philippines/documents/press_corner/renewable_energy_mechanisms_maniego_en.pdf, (accessed 5 Sep. 2013).
92. Energy Regulatory Commission, *supra* n. 82, Rule 7(a).
93. *Ibid.*

*Table 5.1 Technology Type, Installation Target, FIT Rate and Degression Rate*

| Technology/Renewable Energy Source | Installation Target | FIT Rate (Philippine Peso/Kilowatt-Hour) | Degression Rate |
|---|---|---|---|
| Wind | 200 MW | 8.53 | 0.5% after year 2 from effectivity of FIT |
| Biomass | 250 MW | 6.63 | 0.5% after year 2 from effectivity of FIT |
| Solar | 50 MW | 9.68 | 6% after year 1 from effectivity of FIT |
| Hydro | 250 MW | 5.90 | 0.5% after year 2 from effectivity of FIT |

Eligible renewable energy developers can avail of the approved FIT rates upon effectivity of the FIT Allowance as determined by the ERC,[94] which is still being awaited as of this writing.

### [D] Cost-Sharing, Settlement, and FIT Adjustments

The cost of the FIT provided to eligible renewable energy power plants is shared among electricity distribution or transmission network consumers in part through a uniform charge in their electricity bills.[95] Notably, there is no explicit reference that the government provides any subsidy or budgetary allocation in such a cost-sharing arrangement. Upon application by the government-owned National Transmission Corporation, the ERC sets annually the FIT allowance, which appears as a separate uniform charge in the transmission billing statement of the National Grid Corporation of the Philippines in case of consumers directly connected to its system and in the distribution billing statement of distribution utilities in regard to consumers directly connected to their respective system.[96] However, the National Transmission Corporation may immediately file for the adjustment of the FIT allowance if the Working Capital Allowance falls by more than 50% in any quarter of operation of the FIT.[97]

For purposes of the implementation of the FIT, the following factors must be taken into account: (a) the forecasted annual required revenue of eligible renewable energy power plants; (b) the National Transmission Corporation's administration cost; (c) the forecasted annual electricity rates; and (d) such other relevant factors to ensure

---

94. Ibid.
95. Energy Regulatory Commission, *supra* n. 82, Rule 2.4.
96. Ibid., Rule 2.6.
97. Energy Regulatory Commission, Resolution No. 15, Series of 2012, 2.5.

that no stakeholder is allocated with additional risks.[98] The ERC is mandated to adjust the FIT on annual basis for the entire period of its applicability to allow pass-through of local inflation (Philippine Consumer Price Index) and foreign exchange rate variations (Philippine Peso and US Dollar exchange rate).[99] This means that adjustment of the FIT on such instances will apply to eligible renewable energy plants entitled to the FIT for the entire duration of the FIT.

Relevantly, the National Transmission Corporation needs to ensure that the FIT allowance is sufficiently funded to regularly pay all eligible renewable energy power plants, including sufficient allowance for the working capital requirements in case of customer default or delay in collecting and remitting the FIT allowance proceeds.[100] To this end, the ERC is empowered to impose appropriate penalties to minimize such a risk, including the imposition of a 20% penalty surcharge, plus monthly interests on the unpaid amounts. Also, the National Transmission Corporation is allowed to apply for disconnection of any erring party from the grid.[101]

### [E] Priority Connection, Purchase, and Transmission

Power generated from eligible renewable energy power plants enjoy priority connection to the transmission and distribution system subject to compliance with the pertinent standards and rules issued by the ERC.[102] Also, eligible renewable energy power plants are given the priority to inject to the network they are connected and paid the FIT through the National Transmission Corporation based on actual metered deliveries by all on-grid electricity consumers.[103] Unfortunately, the standards and rules for connection, purchase, and transmission remain pending. As such, the cost sharing rule on grid connection and expansion is still unclear.

### [F] Administration and Review

As the entity designated to settle the FIT, the National Transmission Corporation is authorized to: (a) collect information for all renewable energy injections in any transmission or distribution network in the Philippines; (b) audit metering; (c) calculate payments for each eligible renewable energy power plants; (d) collect and make payments; and (e) enter into a Renewable Energy Purchase Agreement with eligible renewable energy power plants.[104] After due proceedings, the ERC shall issue a pro-forma Renewable Energy Purchase Agreement.[105] Any agreement entered into by the National Transmission Corporation and an eligible renewable energy power plant

---

98. *Ibid.*
99. Energy Regulatory Commission, *supra* n. 82, Rule 2.10.
100. Energy Regulatory Commission, *supra* n. 97.
101. *Ibid.*
102. Energy Regulatory Commission *supra* n. 82 Rule 2.7.
103. Energy Regulatory Commission, *supra* n. 97.
104. *Ibid.*, 6.
105. *Ibid.*

in conformity with the pro-forma Renewable Energy Purchase Agreement is deemed approved by the ERC.[106]

The NREB is tasked to regularly monitor and review the development of renewable energy generation and the impact of the FIT.[107] It is also directed to submit a report to the ERC within three years and every two years thereafter.[108] Moreover, the ERC may review and adjust the FIT in case: (a) the installation target per technology is achieved; (b) the installation target per technology is not achieved within the period targeted; (c) there are significant changes to costs or when more accurate data become available that will allow the NREB to calculate the FIT based on such methodology; and (d) other analogous circumstances that justify review and adjustment of the FIT.[109] Although the NREB is directed to ensure that the setting of the FIT and installation targets per technology are consistent with the RPS,[110] it is observed that the approved FIT rates and installation targets per technology have been issued by the ERC without the RPS Scheme being in place.

## §5.04 THE PROPOSED RPS SCHEME

As previously mentioned, the rules governing the establishment of the RPS Scheme are still pending issuance by the DoE. For purposes, however, of identifying the likely implementation arrangement for the RPS Scheme, the following salient features of the proposed 2013 RPS rules are provided.

### [A] Eligible Renewable Energy Technologies

Under the proposed RPS rules, the following renewable energy sources shall be eligible for compliance:

(1) Biomass.
(2) Waste to energy technology.
(3) Wind energy.
(4) Solar energy.
(5) Run-of-river hydro.
(6) Ocean energy.
(7) Hybrid systems.
(8) Impounding hydro sources.
(9) Geothermal energy.
(10) Other renewable energy technologies that may later be identified by the DoE through a separate issuance, upon recommendation of the NREB.

---

106. *Ibid.*
107. Energy Regulatory Commission, *supra* n. 82, Rule 7.
108. *Ibid.*
109. *Ibid.*
110. *Ibid.*

Pertinently, generation from the following renewable energy facilities are deemed eligible for compliance with the RPS:

(1) New installations targeted under the NREP.
(2) Embedded renewable power generating facilities.
(3) Existing renewable energy power generating facilities covered in the baseline RPS calculation.
(4) Incremental capacity resulting from the expansion of an existing renewable energy power generating facility.
(5) Incremental capacity resulting from the upgrading of an existing renewable energy power generating facility that includes retrofitting, refurbishing, or re-powering.
(6) New capacities resulting from a change in the technology, that is, from a non-renewable energy to renewable energy power generating facility such as a coal plant that is modified to use agricultural wastes as fuel.
(7) Mothballed renewable energy power generating facilities that are restored into operation.
(8) Other renewable energy technologies that may later be identified by the DoE through a separation issuance upon recommendation of the NREB.

### [B] Mandated Participant

Per the proposed RPS rules, the following entities are mandated to comply with the RPS:

(1) All distribution utilities for all its existing customers and subsequently, upon commencement of Retail Competition and Open Access (RCOA), for its captive customers.
(2) All licensed Retail Electricity Suppliers (RES) for the contestable market, that is, those end-users who have a choice of supplier of electricity, upon commencement of RCOA.
(3) All local RES upon commencement of RCOA.
(4) Any supplier of last resort, that is, a regulated entity designated by the ERC to serve end-users in the contestable market following a last resort supply event, as may be identified upon commencement of RCOA.
(5) Generating companies only to the extent of their actual supply to their directly connected customers.
(6) Entities duly authorized to operate within economic zones.
(7) Other entities that may be recommended by the NREB and approved by the DoE.

However, the DoE may, on any given year, exempt a Mandated Participant from compliance with the annual RPS requirement under any of the following conditions:

(1) Inadequate supply of the eligible renewable energy sources to meet the annual requirement.
(2) Unavailable capacity at both the transmission and relevant distribution network to transport the eligible renewable energy source generation to the grid.
(3) Occurrence or existence of force majeure affecting or preventing the Mandated Participant from complying with the annual requirements.
(4) Such other consideration or condition as may be determined in the implementation of the REM.

[C]   **Compliance Mechanism**

In complying with the RPS, the Mandated Participant shall use any one, a combination, or all of the following:

(1) Allocation from the System Operator currently the National Grid Corporation of the Philippines pursuant to the FIT Rules.
(2) Renewable energy generation allocated by the System Operator pursuant to the FIT Rules shall be used for compliance purposes and cannot be traded.
(3) Generation from embedded renewable energy power generating facilities, duly certified by the DoE and issued a Certificate of Compliance by the ERC.
(4) Generation from an eligible renewable energy power generating facility with a Power Supply Agreement duly approved by the ERC.
(5) A Renewable energy certificate acquired from the REM where the ownership and value per unit shall be defined by the DoE in a separate issuance.
(6) Any generation from net-metering arrangements.

Upon approval by the DoE, any Mandated Participant may resort to the REC Shortfall Payment (RSP) mechanism to comply with its RPS requirement. This allows a Mandated Participant to pay a pre-determined amount (per kilowatt-hour) to be set by the DoE, which shall not be less than double of the prevailing market price of the RECs. However, the RSP mechanism can be resorted to only under the following instances:

(1) All of the compliance mechanisms provided are no longer available upon certification by the DoE.
(2) There are no sufficient or available RECs in the REM upon certification by the RER.
(3) The reason for non-compliance does not fall within the exemptions provided under the RPS rules and guidelines.

Any person or entity found not complying with the minimum RPS requirement shall be subject to administrative and criminal sanctions.

## [D] Establishment of the Renewable Energy Market and the Renewable Energy Registrar

The DoE shall establish the REM to facilitate the issuance, commercialization and verify compliance with the annual RPS requirement. As part of the REM, the PEMC, under the DoE's supervision, shall establish the RER and shall issue, keep and verify renewable energy certificates corresponding to energy generated from eligible renewable energy power generating facilities. In promulgating the rules and guidelines governing the REM and RER, the following principles will be considered:

(1) The RER will issue one certificate per megawatt-hour of generation produced from a registered generating unit.
(2) The registration shall be designed so a renewable energy certificate can be claimed only once.
(3) All Mandated RPS Participants shall register with the RER their individual RPS Compliance Accounts.
(4) Excess renewable energy certificates of the Mandated Participant can be traded.
(5) A REC shall be valid for three years and can be banked only during its validity.
(6) The Mandated Participant may be assessed periodically with corresponding penalties for non-compliance with the RPS requirement consistent with the validity of the REC.
(7) A Mandated Participant will prove compliance with the RPS by having the proper quantity of RECs in their RPS Compliance Account in the RER.
(8) During the first three years of the RPS program, the DoE shall review the REM rules for possible revisions based on the rate of compliance of the Mandated Participant, REC market activity and general success in meeting the RPS goals.
(9) A transaction fee may be imposed by PEMC for transactions undertaken in the REM and RER subject to the setting of operational charges to be approved by the ERC.

## [E] Minimum Annual RPS Requirement and Annual Increment

The minimum annual target per grid is equal to the sum of the minimum requirement of all Mandated Participants in the said grid. The minimum annual RPS requirement computed for each Mandated Participant serves as its baseline level upon implementation of the RPS and shall be calculated in accordance with the following formula:

Baseline RPS = Min_RPS X E
Min_RPS = Ave_RE or $RE_n$, whichever is lower
AVE_RE = $(RE_n + RE_{n-1} + RE_{n-2} + RE_{n-3} + RE_{n-4})/5$
*Where*:
Baseline RPS = the mandated renewable energy generation for the base year n
Min_RPS = minimum RPS requirement in %

E = total supply portfolio (purchased + embedded generation) to meet the total energy requirement of a Mandated Participant

AVE_RE = 5-year average share of renewable energy to total supply portfolio E

i = refers to year

n = base year

The DoE may apply normalization procedures in any year where the renewable energy share to the total portfolio of a Mandated Participant is deemed to be significantly impacted by a non-recurring situation, which is not attributable to the Mandated Participant. The annual increment in RPS level is initially set at 1% applied to the actual total supply portfolio of the Mandated Participant in the previous year, and may be increased in subsequent years where there is foreseen increase in renewable energy generation. This will determine the number of RECs needed by a Mandated Participant for the current year. The RPS will be reviewed by the DoE once every two years or as may be necessary in consultation with stakeholders.

## §5.05 THE IMPLEMENTATION CHALLENGES TO THE FIT SCHEME

The FIT Scheme is relatively at an early stage of implementation, which makes it difficult to empirically assess the effectiveness of the various design elements of the FIT towards achieving the policy objectives enunciated in the REA. Also, it is challenging to predict the impact of the FIT together with the RPS without seeing how both schemes actually work when fully implemented. However, there are early indications that similar to the FIT Scheme, the RPS rules as currently drafted will require separate issuances to effectively implement the entire RPS Scheme. Again, this will cause delays in its implementation. For example, separate rules for the REM, RER and the REC (ownership and value per unit) are still needed post issuance of the initial RPS rules. Alarmingly, it has been pointed out that the implementation delays put in jeopardy more than USD 2.5 billion worth of potential renewable energy investments in the country.[111] So far, initial drawbacks are already seen in the implementation of the FIT Scheme. These are identified in order to contextualize the challenges ahead for the FIT as a policy mechanism and the REA as a whole to accelerate development of emerging renewable energy technologies in the country.

### [A] Concern on a Customer-Based FIT

One plausible reason that delayed the implementation of the FIT regime in the country is the apprehension that it is an enforced customer-based subsidy mechanism that will arguably lead to an increase in electricity prices.[112] According to the WWF, this design

---

111. WWF Report 2013, *Meeting Renewable Energy Targets: Global lessons from the road to implementation*, http://awsassets.panda.org/downloads/meeting_renewable_energy_targets_low_res_.pdf, 61, (accessed 7 Sep. 2013).
112. European Renewable Energy Council and Greenpeace, *Energy Revolution: A sustainable world energy outlook* (2010) 3rd edition, http://www.greenpeace.org/seasia/ph/Global/international/publications/climate/2010/fullreport.pdf, 21, (accessed 6 Sep. 2013).

feature of the FIT raises payment distribution and equity concerns as it did in the Philippines.[113] Despite the strong opposition from various stakeholders, the FIT has the potential to lower the cost of electricity in the Wholesale Electricity Spot Market (WESM). To understand this proposition, the principle behind the WESM in the Philippines needs to be revisited.

Prior to WESM, the Philippines suffered prolonged power outages[114] due to poorly maintained, inadequate and outdated electric power infrastructure owned by the government-owned National Power Corporation. The Electric Power Industry Reform Act of 2001 introduced WESM as a market-based competitive bidding mechanism to attract new power plant generators and to create additional capacity by 'matching' supply and demand in the market.[115] This works by driving competition among power plant generators, which sell electricity to off-takers at a market price during peak and off-peak demand times. The simple notion is that the more expensive electricity will not be taken by utility companies, if a cheaper one is available in the spot market. However, it plays out differently if supply and demand does not 'match' as what currently prevails in WESM, that is, demand is higher than supply.

Due to the merit-order effect prevailing in WESM where the price for all bidders is set by the last and highest bid offer, the high-demand-low–supply situation benefits sellers, particularly during peak demand times.[116] As a result, electricity price in the spot market is higher due to opportunistic and predatory pricing among suppliers,[117] which is then passed off to consumers in the form of higher electricity bills. It will be noted that the electricity being sold in WESM are primarily generated from conventional power plants that are fueled by imported coal and diesel, which makes electricity pricing vulnerable to the vicissitudes of the international market. This leads us to ask the following question: 'How can renewable energy and the FIT lower the cost of electricity in the spot market?'

In the past, renewable energy technologies were admittedly expensive to make and develop. But recent developments indicate that renewable energy technologies are increasingly becoming cost-competitive. A 2013 report by the International Renewable Energy Agency reveals that the cost of renewable energy technologies such as solar has been on the downtrend for quite some time[118] with the price of solar panels dropping by as much as 48.4% in 2011.[119] This is attributed to the high learning rates in solar photovoltaic technology and its rapid deployment globally.[120] There is even unguarded optimism that the price of electricity generated from solar power technologies will

---

113. WWF Report 2013, *supra* n. 111, 17.
114. *Ibid.*
115. *Ibid.*
116. *Ibid.*, 38–39.
117. *Ibid.*
118. International Renewable Energy Agency, *Renewable Power Generation Costs: An overview* (2013), http://www.irena.org/DocumentDownloads/Publications/Renewable_Power_Generation_Costs.pdf, 4, (accessed 5 Sep. 2013).
119. Deutsche Gesselschaft fur Internationale and Federal Ministry of Economics and Technology, *supra* n. 24, 5.
120. International Renewable Energy Agency, *supra* n. 118, 15.

reach grid parity or the same price as the electricity presently being distributed in the grid sooner than later.[121]

From a cost per kilowatt-hour of electricity generated basis, however, wind energy is identified as one of the most cost-effective renewable technologies currently available in the global market.[122] Cost reductions in wind energy systems are predicted in the coming years as increased competition from suppliers and improvements in the supply chain drive down prices.[123] Considering that the Philippines is endowed with low-cost feedstock for biomass in identified locations across the country, biomass-generated electricity can also be a least-cost solution for providing power to some rural communities.

With a positive outlook on cost reduction for renewable energy technologies and the full implementation of the REA, there is significant potential to bring down the cost of electricity in WESM that will ultimately redound to the benefit consumers. Under a FIT regime, electricity generated from emerging renewable energy sources has priority connection and transmission to the grid aside from being purchased under a guaranteed fixed price for a maximum of 20 years.[124] In effect, renewable energy facilities are deemed 'unscheduled' generators, who do not have to bid and offer a price in WESM, in the same way conventional power plants do.[125] This allows emerging renewables under a FIT regime to significantly supply electricity for priority dispatch – as much as 70% of 2011 WESM sales – which in turn, constrain conventional power producers to compete for the remaining demand by offering lower prices, especially during peak demand times.[126] As the NREB Chairperson explains, a FIT regime provides relative predictability and stability to electricity rates throughout its duration without the 'pass through costing' of fossil fuel power plants.[127] Evidently, the FIT lies at the heart of the implementation of the REA.[128]

### [B]   Issue on Fit Entitlement

Any semblance of implementation delay or lack of decisive action on the part of the government to fully operationalize FIT together with the RPS and the other incentive schemes under the REA will have deep repercussions in the drive towards energy resiliency and sustainable energy development in the Philippines. Recent shifting policy pronouncements on the entitlement to the FIT are sending the wrong signals to the public and the investing community. From the previous policy stance of giving

---

121. *Ibid.*
122. International Renewable Energy Agency, *Renewable technologies: Cost analysis series* (2012), http://www.irena.org/DocumentDownloads/Publications/RE_Technologies_Cost_Analysis-WIND_POWER.pdf, 18, (accessed 5 Sep. 2013).
123. *Ibid.*, 41; 47–50.
124. Energy Regulatory Commission, *supra* n. 82, Rule 4.
125. Greenpeace, *supra* n. 16, 40.
126. *Ibid.*
127. Pete Maniego Jr., *Status of the RE Mechanisms: Sharing Experiences on RE Promotion* (27 Feb. 2012), http://eeas.europa.eu/delegations/philippines/documents/press_corner/renewable_energy_mechanisms_maniego_en.pdf, (accessed 5 Sep. 2013).
128. Dean Tony La Vina, *Our energy choices*, Eagle Eyes, Manila Standard (3 Jul. 2012).

conditional FIT entitlement guarantees to projects at the pre-construction stage, the DoE announced that FIT entitlement can only be endorsed at the post-construction phase.[129] In particular, section 6(g) of DC 2013-05-0009 issued by the DoE on 28 May 2013 clearly stipulates that only 'RE developers holding Certificate of Confirmation of Commerciality shall be issued a [Certificate of Endorsement] COE for FIT eligibility.' This means that the DoE needs to be satisfied first that a renewable energy plant has been successfully commissioned before a COE for FIT eligibility is issued.[130] Effectively, the administrative issuance changes the investment risk consideration for a renewable energy project and is seen by some sectors as favoring the big players or firms with strong balance sheets.[131] It also projects weakened political support for renewable energy development and an impression to preserve the status quo as long as possible, i.e., keeping renewable energy 'marginalised by distortions in the world's electricity markets created by decades of massive financial, political, and structural support to conventional power technologies.'[132]

[C]   FIT Uncertainty upon Full Subscription of Installation Target

Another area of concern that relates to the FIT Scheme is the conservative installation target set initially for the emerging renewable energy technologies eligible to avail the FIT. The Table 5.2 below shows the gap between the Department of Energy-approved installation target and those provided by renewable energy developers, the NREB and the NREP:

Table 5.2   Comparative RE Installation Targets

| Technology | Proposed by RE Developers (2010)[133] | Approved by NREB (2011)[134] | Approved by the Department of Energy (2011)[135] | 2012 NREP Indicative Target (2012–2015)[136] |
|---|---|---|---|---|
| Wind | 710 | 220 | 200 | 1,048 |
| Solar | 542 | 100 | 50 | 269 |
| Run-of-River Hydro | 131 | 250 | 250 | 343.3 |
| Biomass | 416 | 250 | 250 | 276.7 |
| Total | 1,799 | 820 | 750 | 1,937 |

---

129. Myrna Velasco, *supra* n. 63.
130. Department of Energy, Circular No. DC 2013-05-0009, s. 6(e), http://www.doe.gov.ph/issuances/department-circular/2031-dc2013-05-0009 (accessed 7 Sep. 2013).
131. Myrna Velasco, *supra* n. 63.
132. European Renewable Energy Council and Greenpeace, *supra* n. 112, 16.
133. Pete Maniego Jr, *supra* n. 127.
134. *Ibid.*
135. *Ibid.*
136. Jose Layug, Jr., *The National Renewable Energy Program: The Road Starts Here* (27 Feb. 2012), http://eeas.europa.eu/delegations/philippines/documents/press_corner/national_renewable energy_prog_usec_layug_en.pdf.

The comparatively low installation target has considerable implications with respect to attracting new renewable energy projects, which look at the FIT to lower the cost of capital or offset the higher upfront investment costs. It will be noted that in the event that the installation target is fully subscribed the renewable energy developer may avail the FIT only after the next installation target and FIT regime is approved.[137] While a renewable energy developer has the option to enter into a bilateral agreement with a distribution utility or any off-taker or export the power generation directly to WESM if an installation target has reached full subscription,[138] it removes in the meantime the element of revenue certainty until the renewable energy plant is issued a COE for FIT eligibility under the succeeding installation target and FIT regime. The event of the installation target reaching full subscription appears imminent considering that as of 2011 about 384 renewable energy service contracts equivalent to 6,046 MW of generation capacity were awaiting the DoE's approval.[139] With the complicated process associated with the determination of the next installation target and the FIT regime as witnessed in the past, a lot of uncertainty again ensues that only serves to frustrate and diminish whatever interests renewable energy developers have in the country.

## §5.06 CONCLUSION

While the REA has the enormous potential to accelerate renewable electricity development and transform the electricity sector, the inertia of its slow and weak implementation is making it difficult to create the enabling regulatory environment to overcome the barriers to the rapid deployment of renewable energy technologies in the Philippines. The REA incorporates a mix of policy options, tools, and incentives that are designed to be implemented in a timely, integrated, and cohesive manner. The seemingly fragmented and delayed approach to the implementation of the REA defeats its purpose to increase the deployment of renewable energy technologies in the country through effective and interlinked incentive schemes and programs such as the FIT and RPS.

Even at an early stage, serious drawbacks are already apparent in the way the FIT Scheme is designed and implemented. Concern on policy funding through a customer-based FIT, shifting FIT entitlement criteria and impending full installation target subscription indicate the need for greater flexibility and responsiveness in calibrating the policy mechanisms as often as necessary. The same can be said about the RPS Scheme, which follows a similar implementation path as the FIT; delayed, piecemeal, and fragmented. It is noteworthy that the FIT and RPS Schemes are demand-pull policies that work best at the early deployment and commercialization stage of

---

137. Department of Energy, *supra* n. 130, s. 7(c).
138. *Ibid.*, s. 7(a).
139. WWF Report 2013, *supra* n. 111, 62.

renewable energy technologies.[140] This means that complementarities exist between the FIT and RPS in terms of setting overall deployment targets and incentivizing renewable energy projects together to increase renewable energy installed capacities. Accordingly, the FIT without the RPS, including the other incentive schemes and programs, spawn implementation gaps or deficits that only serve to diminish the core and effectiveness of the REA.

Evidently, there is a need to exhibit strong political will to effectively implement not only the FIT and RPS, but also the other complementary policy mechanisms in the REA. The resistance encountered when the FIT Rules were issued demonstrates the need for continuous engagement of the various stakeholders in the implementation process. While there still appears to be considerable interests from renewable energy developers,[141] the Philippines cannot continue to risk losing potential investment opportunities due to unwarranted delays in the implementation of the REA. As the World Energy Council emphasizes, significant work lies ahead to overcome the barriers and concerns to effective renewable energy policy implementation at both the national and local levels.[142] Admittedly, the law will not see its full fruition until the REA is effectively implemented in its entirety beginning with the FIT and RPS.

---

140. Jenna Goodward, Alexander Perera, Nicholas Bianco and Christina Heshmatpour, 'Is the Fit Right? Considering Technological Maturity in Designing Renewable Energy Policy', *WRI Issue Brief* (2011), http://pdf.wri.org/is_the_fit_right.pdf, 6-10, (accessed 6 Sep. 2013).
141. WWF Report 2013, *supra* n. 111, 63.
142. World Energy Council, *Pursuing sustainability: 2010 Assessment of country energy and climate policies* (2010), 19.

CHAPTER 6
# FIT and Its Implementation in Thailand: Legal Measures, Implementation, Challenges, and Solutions

*Robert Brian Smith, Nucharee Nuchkoom Smith & Darryl Robert Smith*

## §6.01 INTRODUCTION

### [A] Overview[1]

Thailand operates a large scale integrated power system. It has a well-developed electricity network and a high per capita energy demand in comparison with its South East Asian neighbors. Malaysia is the only neighbor with a higher degree of electrification and per capita demand.[2] According to the latest World Bank data available (2010), Thailand has a population of around 68 million with 99.3% having access to electricity consuming 2,243 kWh per capita per annum.[3] The electricity consumption in 2010 was 149.32 billion kWh[4] with total energy consumption being 117.43 Mtoe.[5,6]

---

1. One of the main challenges is preparing this paper has been the dearth of available literature in either Thai or English. Government websites provide limited information and that which is available is often a PowerPoint presentation given by senior staff at an international symposium. It is the hope of the authors that this paper assists the reader in better understanding the significant advances being made by the Kingdom of Thailand in the utilization of renewable energy and reduction of greenhouse emissions.
2. Lutz Weischer, Pioneering Renewable Energy Options: Thailand takes up the Challenge, in Inside Stories on Climate Compatible Development, Climate & Development Knowledge Network, May 2013, 1.
3. World Bank, World Development Indicators, World Database, http:/databank.worldbank.org/ (accessed 21 Jun. 2013).
4. Sopitsuda Tongsopit & Chris Greacen, Thailand's Renewable Energy Policy: FiTs and Opportunities for International Support, (Berkeley: University of California, Berkeley Renewable and Appropriate Energy Laboratory, 31 May 2012), www.palangthai.org/docs/ThailandFiTtongsopit&greacen.pdf (accessed 1 May 2013).

Thailand proclaimed the Energy Conservation and Promotion Act in 1992. Electricity in Thailand has been a regulated energy source since 1993.[7] As a result Thailand has, over the years, initiated a number of energy conservation measures including measures for energy conservation in buildings[8] and factories.[9]

The Thailand electricity sector operates under an 'Enhanced Single Buyer' model.[10] The state-owned Electricity Generating Authority of Thailand (EGAT) operates under its own Act[11] and is the own/operator of around 48% of electric generation capacity, with the remaining capacity generally supplied by private operators.[12] Transmission and distribution are split and operated by different government utilities.

EGAT operates the high voltage transmission network whilst the distribution network is operated by the Metropolitan Electricity Authority (MEA) in the Bangkok market and by the Provincial Electricity Authority (PEA) in the provinces, with the Bangkok market constituting 30% of the total market.[13]

In addition to the generation capacity of EGAT of about 48%, Independent Power Producers (IPP) supply around 38% of the power requirements, Small Power Producers (SPP) produce 7%, with imports supplying around 7%; imported power is mainly produced at hydroelectric plants in Laos, with some power also purchased from Malaysia.[14]

Weischer has identified a number of challenges facing the Thai power sector, namely:

- Thailand is dependent on natural gas for over 70% of its electricity generation and imports almost 25% of its natural gas supply.
- Whilst there is disagreement over future electricity demand predictions, there is general agreement that additional capacity will be required in coming years.
- Thailand recognizes that it needs to reduce pollution and greenhouse gas emissions and has set an objective of being a low carbon society.
- As the power sector was responsible for 42% of greenhouse gas emissions in 2011, it will have to make a significant contribution to the reduction effort.[15]

It should be noted that the levels of greenhouse gas emissions as reported would have been much far higher if coal was used as the predominant fuel source with its higher greenhouse gas emissions.

---

5. International Energy Agency, World Energy Statistics, 2012, 56, http://www.iea.org/publications/freepublications/publication/kwes.pdf (accessed 29 Jun. 2013).
6. Tonne of oil equivalent (toe) is the International Energy Agency/OECD unit of energy and is the amount of energy released by burning one tonne of oil.
7. The Royal Decree on Regulated Energy, BE 2536 (1993).
8. The Royal Decree on Designated Building, BE 2538 (1993).
9. The Royal Decree on Designated Factory, BE 2540 (1995).
10. Tongsopit & Greacen, *supra* n. 4, at 2.
11. Electricity Generating Authority of Thailand Act, BE 2511 (2006).
12. Pallapa Ruangrong, Energy and Regulatory Overview of Thailand, PowerPoint Presentation to Asia Pacific Energy Regulators' Forum (APER), 1 Aug. 2012, Washington D.C, USA.
13. *Ibid.*
14. *Ibid.*
15. Weischer, *supra* n. 2, at 1-4.

As a result, Thailand decided to develop a series of Development Plans culminating in the 10-year Alternative Energy Development Plan (2012–2021), which set a target of 25% of total energy consumption by 2021 to be provided by renewable energy, with 10% of electricity consumption being met by renewable energy.[16]

Unlike other markets where renewable energy is being promoted primarily on energy security grounds, Thailand's growth is commercially motivated and is driven by financial incentives and supporting policies.[17] In other words, Thailand sees a commercial advantage in producing green energy from alternative energy sources. Not only does it reduce reliance on potentially costly imports, it fosters economic growth by development of new industries and the production of renewable energy sources from agricultural products.

Of the six most developed ASEAN countries, Thailand is seen as having the highest renewable energy targets, the highest level of financial incentives as well as the highest level of non-financial incentives.[18] However, Thailand, like Malaysia, is perceived to have medium importance issues/risks associated with the administrative/regulatory environment; market related issues, technical and infrastructure issues; and finally in the area of socio-cultural issues.[19]

The energy industry in Thailand operates under a mix of legislative and administrative requirements. It is primarily governed by the Energy Industry Act BE 2550 (2007).

### [B] Energy Industry Act BE 2550 (2007)

The Energy Industry Act BE 2550 (2007) was enacted to apply to the operation of the energy industry throughout the Kingdom (section 4) and its application is restricted to electricity and natural gas (section 5).[20]

The objectives of the Act as set out in section 7 include procuring sufficient energy to adequately meet the demand in a sustainable manner; promoting competition in the energy industry; promoting economical, efficient and worthwhile use of energy whilst considering the environmental impacts, and increasing the economic competitive edge of the country; encouraging increased participation of the local communities and developing education programs to promote energy conservation; and promoting the use of renewable energy.[21]

---

16. DEDE AEDP 2012-2021 presented on DEDE Website, http://www.dede.go.th/dede/images/stories/aedp25.pdf in Thai version. English by Dr. Renu Cheokul (April 2012) (Accessed 1 May 2013).
17. Ipsos Business Consulting, Meeting the Energy Challenge in South East Asia: A Paper on Renewable Energy, July 2012. http://w3.ipsos.com/businessconsulting/insights/whitepaper/docs/A_Paper_on_Renewable_Energy_in_South_East_Asia_July_2012.pdf (accessed 26 Jun. 2013).
18. Samantha Ölz and Milou Beerepoot, Deploying Renewable in South East Asia: Trends and Potentials. (Paris: International Energy Agency, 2010), 136.
19. Ibid., 137.
20. Energy Industry Act 2550 (2007) (Unofficial Translation), www.eppo.go.th/admin/cab/law/energy_industry_act-2007.pdf, (accessed 23 Jun. 2013).
21. Ibid.

The definition of the term 'renewable energy' refers to section 4 of the National Energy Policy Council Act BE 2535 (1992) and includes 'energy obtained from wood, firewood, paddy husk, bagasse, biomass, hydropower, solar power, geothermal power, wind power, and waves and tides' whilst non-renewable energy includes 'energy obtained from coal, oil shale, tar sands, crude oil, oil, natural gas, and nuclear power'.[22]

Under the Energy Act, the role of the government, as set out in section 8, is to establish fundamental policy guidelines on the energy industry. The role of the government is to procure sufficient energy to adequately meet the demand in a sustainable manner; promoting competition in the energy industry; promoting economical, efficient, and worthwhile use of energy whilst considering the environmental impacts, and increasing the economic competitive edge of the country; encouraging increased participation of the local communities and developing education programs to promote energy conservation.[23]

It might be noted that whilst renewable energy is included in section 7, it is not included in section 8. In addition, following a failed attempt at power industry privatization, this section states:

> The government will be responsible for electricity network system operation, electricity system operation and hydropower plants – with the Electricity Generating Authority of Thailand being the operator of the electricity transmission system, the Metropolitan Electricity Authority and the Provincial Electricity Authority being the operators of the electricity distribution systems – including retention of appropriate reasonable proportion of electricity generation capacity of state-owned electricity industry.[24]

Moreover, the Act sets out the authority and duties of the Minister of Energy in relation to his administration. The powers and duties of the Minister as set out in section 9 include:

- proposing to the Cabinet policy on energy industry structure;
- proposing to the National Energy Policy Council (NEPC), policies on energy procurement and diversification of fuel sources to ensure efficiency and security of electricity industry;
- considering the power development plan, the investment plan of electricity industry, the natural gas procurement plan and the energy network system expansion plan, for submission to the Cabinet for approval; and
- proposing the NEPC policy on the level of contributions to the Power Development Fund and on the utilization of those funds.[25]

---

22. National Energy Policy Council Act BE 2535 (1992) (Authorized Official Translation), http://www.thailawforum.com/database1/national-energy-act.html, (accessed 24 Jun. 2013).
23. Energy Industry Act 2550, *supra* n. 20.
24. *Ibid.*, s. 8 (5).
25. *Ibid.*, s. 9.

Because of the critical role of energy to the nation's economy an Energy Regulatory Commission (ERC) is established under Division 2 of the Act. The Commission's authorities and duties include:

- regulating energy industry operation;
- imposing measures to ensure security and reliability of electricity system;
- providing opinions on the power development plan, the investment plan of the electricity industry, the natural gas procurement plan, and the energy network system expansion plan for submission to the Minister;
- issuing regulations and announcements and supervising customer service standards and quality;
- issuing regulations or announcements on criteria, method and conditions of the contributions given to the Power Development Fund and the utilization of those funds;
- promote and support study and research on energy industry operation;
- promoting energy awareness;
- promoting economical and efficient use of energy, renewable energy and energy that has minimal impact on environment, with due consideration of efficiency of electricity industry operation and balance of natural resources.[26]

It is the ERC, then, which has prime responsibility for implementing renewable energy policies in the electricity and natural gas industries.

The ERC is then subject to a number of regulations as described in Commentaries of Laws related to Energy Industry, ERC, Volume 1.[27]

This is the legislative framework under which the Feed-in Tariff Scheme operates.

Thailand currently operates an Adder Program and is in the process of moving to a (FIT) Feed-in Tariff Scheme.[28]

## §6.02 FEED-IN TARIFF SCHEME

### [A] The Detailed Design of the Feed-In Tariff Scheme

*[1] Technology Eligibility*

The renewal energy program is administered by the three government electricity utilities (EGAT, MEA, PEA) that purchase electricity generated from renewal energy from:

---

26. *Ibid.*, s. 9.
27. *Ibid.*, Division 2.
28. Tongsopit & Greacen, *supra* n. 4, at 3.

- SPP for generators sized greater than 10 MW and less than 90 MW and operating under the SPP regulations.[29]
- Very Small Power Producers (VSPP) for generators sized less than or equal to 10 MW and operating under the VSPP regulations.[30]

It is considered that the classification should be based on 'registered' capacity and not generating capacity. This, for instance, allows a wind station rated at 60 kW to be registered as 50 kW, on the basis that it is limited to generating 50 kW. But since the fans are bigger, it would potentially generate a better dollar return for the operator given the different tariffs for SPP and VSPP producers. This also allows excess generation capacity to be used by the operator for its own needs. In addition there appears to be no impediment under the Act for an operator to store excess energy and feed it back into the grid later to meet its supply obligations at the FIT rate.

The types of renewable energy eligible for inclusion in the schemes are biomass, biogas from all sources; waste both municipal solid waste (MSW) and non-toxic industrial waste; wind; hydro (mini- and micro-hydro); and solar.[31]

In addition, VSPPs are able to produce power using photovoltaics, sea or ocean waves, and geothermal energy.[32]

On 28 June 2010, the NEPC resolved not to accept any further solar energy projects until there was a review of policy and guidelines (section 13.3).[33] At the same meeting, the Commission agreed in principle to a proposal to introduce a Feed-in Tariff for solar projects that are installed on residential and commercial buildings (section 12.2).[34] The argument being that it will foster energy efficiency by promoting the installation of solar energy on residential and commercial roofs as it reduces the power loss in the system because it is produced and used at the point of installation and does not require a lot of space.[35] At the time of writing this paper in July 2013, the applications for solar projects were still on hold, with the details of the FIT for solar projects on residential and commercial buildings still not released.

---

29. Regulations for the Purchase of Power from Small Producers (1998) revised August 2001, http://cdm.unfccc.int/Projects/DB/DNV-CUK1174895235.33/ReviewInitialComments/Y1IBB4M1S9LY4C9M8ZAM318OS7N3QK (accessed 27 Jun. 2013).
30. Regulations for the Purchase of Power from Very Small Power Producers (for the Generation Using Renewable Energy) (Unofficial Translation), http://www.eppo.go.th/power/vspp-eng/Regulations%20-VSPP%20Renew-10%20MW-eng.pdf (accessed 24 Jun. 2013).
31. Pallapa Ruangrong, Thailand's Power Tariff Structure, PowerPoint Presentation to Regional Energy Regulatory Associations of Emerging Markets, Roundtable Discussion III: Affordability and Customer Issues, 8–9 Apr. 2013, Istanbul, Turkey.
32. Regulations for the Purchase of Power from Very Small Power, *supra* n. 30.
33. Resolution of the National Energy Policy Council. 2/2553 (No. 131.) on Monday, 28 June, 2553 (in Thai), http://www.eppo.go.th/nepc/kpc/kpc-131.htm (accessed 25 Jun. 2013).
34. *Ibid.*
35. *Ibid.*

On 8 February 2013 the NEPC approved the use of biogas from energy crops under the Community Enterprise green energy plants in the form of an Adder for projects that can produce up to 1 MW of power at a rate equivalent to USD 0.15 per unit for a period of 20 years; the plan being to encourage farmers grouped together as communities or cooperatives to grow energy crops.[36]

Provided all environmental clearances are obtained, there appears to be no impediment to the establishment of Community Enterprise green energy plants that provide local power that is used locally and is not purchased by the power utility.

*[2] FIT Duration*

The period of support from the FIT commences on the Commercial Operation Date (COD) and is 7 years for biomass, biogas, waste, and hydro; and 10 years for wind and solar.[37]

As noted above, biogas from energy crops under the Community Enterprise Green Energy plants are supported for a period of 20 years.[38]

*[3] Tariff*

*[a] Tariff Schedule*

The rate for feed-in tariffs in Thailand since 2007 is paid on top of the utilities' avoided costs and is called a Premium Feed-in Tariff or Adder.[39] In 2010, the government approved a plan to switch from a premium-price FIT payment to a fixed-price FIT.[40] Studies to determine the rates for each type of renewable energy are still being considered at the time of this paper. The current Adder rates are shown in Table 6.1. Contracts are in Thai baht and have been converted here for comparison purposes.

---

36. Resolution of the National Energy Policy Council. No. 1/2556 (No. 144.) (in Thai) http://www.eppo.go.th/nepc/kpc/kpc-144.htm (accessed 25 Jun. 2013).
37. Tongsopit & Greacen, *supra* note 4 at 8, and Ruangrong, *supra* n. 31.
38. Resolution of the National Energy Policy Council. No. 1/2556 *supra* n. 36.
39. Fixed Feed-in Tariff is the determination of a purchase rate for electricity generated from renewable energy sources at a certain constant level which is independent from the fluctuation of market price for electricity throughout the support duration. With a Premium Feed-in-Tariff. a premium or Adder is an additional rate on top of the market price for electricity. Therefore, the purchase price for electricity generated from renewable energy sources will fluctuate in line with the market price.
40. Pallapa Ruangrong, Thailand's Power Tariff Structure, PowerPoint Presentation to Regional Energy Regulatory Associations of Emerging Markets, Roundtable Discussion III: Affordability and Customer Issues, 8–9 Apr. 2013, Istanbul. Turkey.

Table 6.1  Thailand's Adder Rates (Exchange Rate USD 1 = THB 30)[41]

| Type of Renewable Energy | Former Adder 2009 ($/kWh) | Adder as of 2010 ($/kWh) | Additional for Diesel Substitution ($/kWh) | Additional for RE Generators in the Three Most Southern Provinces ($/kWh) |
|---|---|---|---|---|
| **Biomass** | | | | |
| Installed Capacity ≤ 1 MW | 0.017 | 0.017 | 0.033 | 0.033 |
| Installed Capacity > 1 MW | 0.010 | 0.010 | 0.033 | 0.033 |
| **Biogas (All Sources)** | | | | |
| Installed Capacity ≤ 1 MW | 0.017 | 0.017 | 0.033 | 0.033 |
| Installed Capacity > 1 MW | 0.010 | 0.010 | 0.033 | 0.033 |
| **Waste (MSW and Non-toxic Industrial Waste)** | | | | |
| Fertilizer/Landfill | 0.083 | 0.083 | 0.033 | 0.033 |
| Thermal Process | 0.117 | 0.117 | 0.033 | 0.033 |
| **Wind** | | | | |
| Installed Capacity ≤ 50 kW | 0.150 | 0.150 | 0.050 | 0.050 |
| Installed Capacity > 50 kW | 0.117 | 0.117 | 0.050 | 0.050 |
| **Hydro (Mini/Micro Hydro)** | | | | |
| 50 kW < Installed Capacity < 200 kW | 0.027 | 0.027 | 0.033 | 0.033 |
| Installed Capacity ≤ 50 kW | 0.050 | 0.050 | 0.033 | 0.033 |
| **Solar** | | | | |
| | 0.267 | 0.267 | 0.217 | 0.050 |

The 2009 tariff was set by a Cabinet resolution of 24 March 2009 and was valid from that date.[42] As will be seen later in this paper, the tariff for solar energy generators was reduced in 2010. This was due to the reduction on capital costs due to delay in approved suppliers entering the supply chain as the prices of solar cells were falling significantly. Energy suppliers determined that their windfall profit would be greater the longer they waited.[43] Suppliers delaying generation projects had an improved financial outcome at a negative environmental outcome. The revised rate applied to projects that had not been accepted by the power utilities by that date. The 2010 rates

---

41. Ibid.
42. Ibid.
43. Energy Policy and Planning Office Annual Report BE 2554 (2010), 130.

apply at the date of this paper in July 2013. This outcome provides a very obvious example of the need for a regression pricing mechanism as is, indeed, proposed for Thailand.

Biogas from energy crops under the Community Enterprise Green Energy Plants in the form of Feed-in Tariff for projects that have the power to sell up to 1 MW is supported at a rate of USD 0.15 per unit for a period of 20 years.[44] It should be noted that this is saleale capacity. Additional generation capacity could be installed and not exported to the grid due to self-use or other consumption.

As was seen in Table 6.1, the current FIT is dependent on both energy type and location. An additional tariff is paid for power generation from the three southern provinces.

*[b] Tariff Degression Mechanism*

On 28 June 2010, the NEPC ordered a review of the renewable energy support program.[45] A Feed-in Tariff is proposed to replace the Adder, but as of July 2013 details are not available.

The only degression mechanism that has been used is the one-off decision to reduce the Adder for solar energy in 2010.

*[c] Tariff Progression Mechanism*

There is no tariff progression mechanism in place, although there is a recognition that the current Adder rates for some forms of renewable energy are inadequate to encourage development of that sector due to uncommercial pay back periods.

**[4] Capacity Cap**

*[a] Soft Cap*

The monthly capacity factor for a SPP must not be less than 0.51 but no more than 1.0 times the agreed Contract amount (MW), except when otherwise requested by the utility. The utility may, however, require the SPP to be able to generate and supply power in accordance with power utility's requirement (but not exceeding the quantity indicated in the contract).

The amount of net power each VSPP dispatches into the distribution system must not exceed 10 MW at any time. The Distribution Utility will, however, consider the

---

44. Resolution of the National Energy Policy Council. No. 1/2556 *supra* n. 36.
45. Wattanapong Karovat, Financial Mechanism for Renewable Energy PowerPoint Presentation to Seminar of Financial Schemes for Renewable Energy Projects, 27 Nov. 2012, Landmark Hotel Bangkok.

capability and security of the distribution system in determining the level of net power acceptable on a case-by-case basis.[46]

### [b] Hard Cap

At the moment the cap is controlled by a 'Cap and Deadline' mechanism.[47] The first phase of the Adder program had a deadline at the end of 2008. When the scheme was revised in March 2009, no deadline was imposed but the NEPC imposed a broad guideline that new project approval would be subject to acceptable cumulative effects on pass-through costs to consumers.[48] Unfortunately, there is no guidance as to when the pass-through cost becomes unacceptable with the utility companies being aware of this eventual ceiling but having no guidance as to when they must stop accepting further applications.[49] This results in the utilities using their own discretion in accepting or rejecting applications.[50]

### [5] Loading Hours (Resources Quality Cap)

#### [a] Small Power Producers

SPPs are required to meet the conditions set out in the SPP Regulations.[51]

SPPs must generate and supply electricity to the Power Utility during the system peak months of March, April, May, June, September and October, and the total hours of electricity production supplied to the utility must be no less than 7,008 hours per year.[52]

For SPPs using waste or residues from agricultural processes or from industrial productions, processes or products derived from these processes, garbage and dendro-thermal sources (such as tree plantations), the annual hours must be not less than 4,672 hours per year.[53] Generation and sales must include the period of March, April, May, and June.[54]

As noted above utility may require the SPP to be able to generate and supply power in accordance with the utilities requirement (but not exceeding the quantity indicated in the contract).[55]

---

46. Regulations for the Purchase of Power from Very Small Power Producers, *supra* n. 30.
47. Tongsopit & Greacen, *supra* n. 4, at 8.
48. Ibid.
49. Ibid.
50. Ibid.
51. Regulations for the Purchase of Power from Small Producers, *supra* n. 25.
52. Ibid.
53. Ibid.
54. Ibid.
55. Ibid.

The quality of electricity generated must be in accordance with the Regulations for the Synchronization of Generators to the System of the particular power utility to ensure the security of the electricity network.[56]

Shut-down for planned maintenance is only allowed to take place during the off-peak months of the system which include the months of January, February, July, August, November, and December and must not exceed 840 hours or 35 days in a 12-month cycle.[57]

In the case of an emergency, the total shut-down time for maintenance during the peak demand period (18.30–21.30 hours) of the peak months should not exceed 30 hours for a 12-month cycle.[58]

These regulations are designed to ensure that planned outages are minimized, and that the forced outage rate is as low as possible.

*[b]   Very Small Power Producers*

Very Small Power Producers are required to meet the conditions set out in the VSPP Regulations.[59]

Unlike in the case of SPP, there is no requirement to supply a minimum amount of power nor is there a requirement to provide power at peak times.

The quality of electricity generated must be in accordance with the Regulations for the Synchronization of Generators to the System of the particular power utility.[60]

*[6]   Cost Sharing and Recovery*

SPPs are responsible for the cost of system interconnection, which includes the costs of the transmission and distribution system of the SPPs and the Public Utility, the meters, the protective devices, and other expenses arising from undertaking purchasing electricity from the SPPs and are also responsible for cost of equipment inspections.[61] They are also required to install protective devices to prevent damage to the system as prescribed in the Regulations Governing Synchronization of Generators and each party is responsible for damage caused by faulty electrical devices or other causes that arise from its own system.[62] The conditions for VSPPs are similar.[63]

---

56. Ibid.
57. Ibid.
58. Ibid.
59. Regulations for the Purchase of Power from Very Small Power Producers, *supra* n. 30.
60. Distribution Utilities' Regulations for Synchronization of Generators with Net Output under 10 MW to the Distribution Utility System, http://www.eppo.go.th/power/vspp-eng/VSPP%20Synchronization%2010%20MW-eng.pdf (accessed 28 Jun. 2013).
61. Regulations for the Purchase of Power from Small Producers, *supra* n. 29, s. G.
62. *Ibid.*, s. N.
63. Regulations for the Purchase of Power from Very Small Power Producers, *supra* n. 30, s. G.

In addition, there are a number of financial incentives available for renewable energy projects. These include a Revolving Fund that provides a loan at a maximum interest rate of 4% for a period of up to seven years and the Energy Conservation Promotion Fund (ESCO), which provides venture capital.[64] Finally Thailand's Board of Investment (BOI) provides tax incentives for renewable energy.[65] These include a tax holiday of up to eight years, exemption or reduction of import duties on solar equipment, and corporate income tax reduction.[66]

### [7] Grid Connection, Usage, and Expansion Rules

Initially, the VSPP program utilities were required to grant projects permission to interconnect to the power network if they met basic safety and power quality standards, provided there was sufficient substation capacity.[67] This all changed in 2010 when more rigorous requirements were implemented.

Prior to the EGAT, the PEA and the MEA entering into a power purchase agreement, the Managing Committee on Power Generation from Renewable Energy Promotion, established by NEPC, has nominated the criteria that must be assessed prior to making a decision on power purchase.[68] The project must have a connecting point which can be easily identified and be well equipped with a certain Scheduled Commercial Operation Date (SCOD). The transmission and/or distribution system must be able to support the electricity purchase according to the SCOD. The project must be technically approved by EGAT and have an appropriate and clear operating plan. [This power is granted to EGAT under section 18, Chapter 2 of the EGAT Act BE 2511]. If biogas or garbage is to be used, the fuel source must be identified. Tyres and other forms of polluting garbage are not to be used. If wind energy is to be used, there must be a declaration that the proponent has rights to use the land.[69]

Once these initial criteria are met another five criteria must be met before a power purchase contract can be signed.[70] The project must have received a confirmation for power purchase from the relevant Electricity Authority and must be technically approved by EGAT. The project must have gone through an examination to ensure its readiness in relation to legal possession of the required land, obtained access to the required capital, possession of the required technology and possess or being in the process of requiring all licenses required by law. The proponent must agree to take responsibility for any system development costs.

---

64. Karovat, *supra* n. 45.
65. *Ibid.*
66. *Ibid.*
67. Tongsopit & Greacen, *supra* n. 4 at 14.
68. Energy Policy and Planning Office Annual Report BE 2554 (2010),125.
69. *Ibid.*
70. *Ibid.*

Finally, the project must have an Environmental Impact Assessment report as required by law and receive approval from the authorized government agency.[71] This is particularity pertinent in Thailand where there are a number of cases of forced resettlements where villages claim that they were not consulted before they were ordered to move. In September 2003, Government officials ordered the residents of four small mountain villages in Lampung Province to move.[72] Villagers claim they were not consulted before the decision to move them was made and were told only that its purpose was to improve their 'development' and when a number of them protested, officials are reported to have told them that they would be punished if they did not move, and if they did, they were reportedly promised comparable land upon resettlement, as well as citizenship cards.[73]

In January 2011, the three power authorities were instructed to issue a notification prohibiting a change in or amendment to Power Purchase Agreements with renewable energy projects applying for a change in the quantity of the power energy offered for sale, relocation of the power plant, or a change in the production technology.[74] EGAT issued such notification dated 14 March 2011.[75]

### [B] The Results of the FIT Implementation

As can be seen in Table 6.2, the quantity of electricity offered by private power producers in areas such as solar energy and biomass far exceeded the target in the 15-year Renewable Energy Development Plan (REDP). The actual volume of electricity distributed into the grid was far below the agreed volume in the Power Purchase Agreements.

This is particularly in the case of solar energy, where projects were clearly delayed to take advantage of the marked drop in the price of solar panels.

Karovat has identified a number of difficulties with the current scheme operating in Thailand and the rationale behind the proposed Feed-in Tariff regime:

- The current Adder is dependent on the global energy pricing; the Feed-in Tariff is not.
- The Adder poses a potential long-term risk to both the developer and the end user; with a revised Feed-in Tariff the risk to all parties is less.

---

71. *Ibid.*, 126.
72. Asia Pacific Forum of National Human Rights Institutions / Brookings Institution – SAIS Project on Internal Displacement National Human Rights Commissions and Internally Displaced Persons Project Visit of a Three Member Team to Thailand, 9–13 Aug. 2004, http://www.oknation.net/blog/print.php?id = 169463 (accessed 29 Jun. 2013).
73. *Ibid.*
74. Chandler & Thong-EK, Wind Energy Development in Thailand, 22 Mar. 2011, http://www.ctlo.com/mediacenter/2011-03-29-MemoreWindEnergyDevelopmentinThailand_(330445_3).pdf (accessed 28 Jun. 2013).
75. *Ibid.*

- The Adder is an up-front subsidy to the developer by providing faster payback whilst imposing an extra burden on the consumer.
- The Adder with its up-front subsidies may promote inefficient technologies whilst the Feed-in Tariff promotes the use of high energy efficient technology.
- The Feed-in Tariff leads to less uncertainty as to the amount of renewable energy that will be available to assist in power development planning.[76]

The significance of Karovat's analysis is shown by the vast difference between what was offered in the Power Purchase Agreements and that which was actually produced. This makes reliable energy planning almost impossible. The potential solar energy producers were no doubt waiting to enhance their windfall profits by building their infrastructure as late as possible to take advantage of the continuing decline in the price of solar panels whilst the Adder rate remained unchanged.

---

76. Karovat, *supra* n. 45.

Table 6.2 Purchased Electricity from Renewable Energy in November 2011 Compared with Target Volume in the REDP[77]

| Fuel Type | Target Volume REDP (MW) | With Actual Transmission to the Grid | | With PPA Already Signed (Waiting for COD) | | With Acceptance to Purchase (Pending PPA Signing) | | Under Consideration for Purchase | |
|---|---|---|---|---|---|---|---|---|---|
| | | Number of Projects | Volume Offered (MW) | Number of Projects | Volume Offered (MW) | Number of Projects | Volume Offered (MW) | Number of Projects | Volume Offered (MW) |
| Solar Energy | 500 | 85 | 65.14 | 438 | 2,048.33 | 65 | 368.41 | 183 | 1,116.85 |
| Biogas | 120 | 70 | 106.93 | 33 | 60.28 | 26 | 41.92 | 47 | 76.00 |
| Biomass | 3,700 | 84 | 704.72 | 220 | 1,564.15 | 50 | 406.69 | 36 | 264.43 |
| Municipal Solid Waste (MSW) | 160 | 12 | 37.33 | 21 | 108.67 | 6 | 7.28 | 8 | 78.45 |
| Hydro Energy | 324 | 6 | 13.28 | 5 | 6.10 | 2 | 0.305 | 1 | 0.03 |
| Wind Energy | 800 | 3 | 0.38 | 17 | 308.28 | 24 | 356.25 | 17 | 488.49 |
| Total | 5,604 | 260 | 027.78 | 734 | 4,095.81 | 173 | 1,180.85 | 292 | 2,024.25 |

77. After Energy Policy and Planning Office Annual Report BE 2554 (2010), 130.

## §6.03 RECENT DISCUSSION OVER FIT AT A POST-FUKUSHIMA AND POST-KYOTO PROTOCOL ERA

### [A] Evaluation

As previously noted, Thailand is a net energy importer. The data from 2011 showing that more than 60% of primary commercial energy demand is derived from imports with oil imports running at 80% and increasing as domestic production is incapable of increasing to meet demand. Around 70% of power generation is dependent on natural gas.

Thailand reacted quickly to the Fukushima Daiichi Nuclear Power Plant disaster as they sensed that this disaster lessened public acceptance and trust in the Thailand's nuclear power project development. On 3 May 2011, Cabinet endorsed a recommendation by the Ministry of Energy to defer SCOD of the first unit of Thailand's first nuclear power project 2020-2023 to allow for detailed review of the Project.[78] The review is to consider safety measures, legislation framework for development and operation of nuclear projects, stakeholder opinion and future involvement in the process as well as development of additional supporting plans and resulted in preparation and expeditious acceptance of the Power Development Plan 2010 Revision 2.[79]

The Power Development Plan also took cognizance of the government policy that is targeting on increasing the share of renewable energy and alternative energy uses by 25% instead of fossil fuels within the next 10 years, by initiating new projects of renewable energy development.[80] At the end of 2030, total capacity of renewable energy is to proposed to be 20,546.3 MW (or 29% of total generating capacity in the power system) consisting of domestic renewable energy of 13,688 MW and renewable energy from neighboring countries of 6,858 MW.[81]

Whilst these figures provide generating capacity they do not reflect the actual supply of power to the network as they are dependent on availability as some sources provide power continuously whilst others like solar power have restricted availability (e.g., about 1,000 hours per year). When calculated on the basis of kWh the anticipated power consumption in 2030 is expected to be 300,380 kWh with renewable energy providing 19,732 kWh (6.57%), hydro providing 6.941 kWh (2.31%) and nuclear 16,046 kWh (5.34%).[82] The proposed sources of renewable energy are solar power, wind power, hydro power (both domestic and from neighboring countries), biomass, biogas, and MSW.[83]

---

78. Ministry of Energy, Energy Policy and Planning Office, *Summary of Thailand Power Development Plan 2012 – 2030.* (PDP2010 Revision 3), 2012, 1, http://www.eppo.go.th/power/PDP2010-r3/PDP2010-Rev3-Cab19Jun2012-E.pdf accessed 19 Jun. 2013.
79. *Ibid.*
80. *Ibid.*
81. *Ibid.*
82. Ruangrong, *supra* n. 12.
83. Ministry of Energy, *supra* n. 78, at 1.

In 2010 the NEPC established the Managing Committee on Power Generation from Renewable Energy Promotion.[84] This has resulted in more stringent policies and has centralized all decisions regarding SPP and VSPP renewable energy projects under the Managing Committee, which has also taken on responsibility for policy design (the purview of the Energy Planning and Policy Office (EPPO)) regulation (the purview of the ERC).[85] Clearly, this is an issue that must be addressed, so that duplication of effort is avoided, and a clear and transparent approval process is established.

Weischer has identified a number of factors to explain why Thailand was able to take up the challenge of renewable energy:

- First, the renewable energy policies were aligned with broader political considerations beyond environmental considerations. In essence they encouraged private participation in the sector yet were small enough not to seem to be threatening the government owned utilities.
- Civil society played a crucial role in the design of the VSPP and the Adder. As there was very limited expertise within the Thai electricity sector, civil society organizations such as Palang Thai[86] brought in overseas exerts on renewable energy and regulatory frameworks that have worked in other countries.
- Thai programs started small and grew over time, essentially providing pilot schemes that showed both regulators and the wider community that the program would work.[87]

The Alternative Energy Development Plan 2012–2021 has also acknowledged the impact of global warming due to greenhouse gases.[88] Although Thailand has not agreed to enforcement, at the moment the Plan acknowledges that Thailand should conduct the renewable energy development and promotion as a measure to reduce the release of greenhouse gases as 'this would be an initial point to step into the Low Carbon Society and be exemplary for the world society to cite Thailand as the country with strong intent in using renewable energy'.[89]

## [B] Challenges

The Alternative Energy Development Plan 2012–2021 has identified new energy resource types for power generation. There are, however, a number of challenges that must be met.[90]

The target for geothermal energy is planned to increase from 350 kW to 1 MW over the period. The identified issues include the lack of domestic geothermal sources with high heating value and the reliance on overseas technologies. They have also

---

84. Tongsopit & Greacen, *supra* n. 4 at 14.
85. *Ibid.*
86. www.palangthai.org/.
87. Weischer, *supra* n. 2 at 4–5.
88. DEDE AEDP 2012-2021 *supra* n. 16.
89. *Ibid.*
90. *Ibid.*

identified the need for community education on the production of electricity using geothermal energy.[91] This recommendation is no doubt partly due to the poor history of consultation with affected communities and their subsequent resettlement in northern Thailand, as outlined earlier. Northern Thailand is the potential source of geothermal energy.

The proposed roadmap involves the development of a map of potential geothermal sources and the identification of appropriate technologies.[92] The preferred option is to adopt technology that utilizes geothermal energy at the prevailing lower temperatures likely to be encountered in Thailand. Following initial studies, it is proposed to evaluate the cost effectiveness of the technology for the geothermal source and its geography. In addition, they propose to assess the impacts on community, environment, and public health from energy production before proceeding.

Another energy target is power generation from wave and tidal current that is not utilized at present but has a target of 2 MW.[93] The major impediment is the lack of data and an assessment on wave and tidal energy potentials. The proposed roadmap requires an acceleration of studies to assess the potential tidal and wave energy sources and appropriate technologies for power generation. The primary sites identified in the Plan are Sarasin Bridge at Phuket on the western coast of peninsular Thailand and the island areas surrounding areas of Koh Samui on the east of the peninsular east. Once the site has been identified it is necessary to undertake a capably assessment of the development potential and readiness preparation to develop a pilot project.

Whether or not Thailand can reach this target is questionable. Worldwide, only tidal barrages, exploiting tidal rise and fall, are a mature technology, and can face environmental controversy; tidal/ocean currents and wave power are still at the demonstration stage whilst temperature and salinity gradient technologies remain at the research and development stage.[94]

One of the sites identified in the plan, namely Sarasin Bridge is known by the authors, and any project at this site will have significant environmental, and social impacts, particularly on the fishing industry and potentially on fragile mangrove areas. The other locations are near tourist islands and if not carefully managed could also have significant impacts.

Finally, the plan includes the use of hydrogen energy and energy storage system.[95] The Plan has identified a number of major problems and barriers including both a low priority given to domestic research and development and a lack continuous of budget support. As a result, energy development will be dependent on overseas technology. There is no current measure to provide incentives in the development and utilization of hydrogen as either an energy source or as an energy storage system. The roadmap includes studying appropriate raw material sources of hydrogen production in Thailand; research and development technologies for domestic production, storage,

---

91. *Ibid.*
92. *Ibid.*
93. *Ibid.*
94. International Energy Agency, Renewable Energy Medium-Term Market Report 2012, 149.
95. *Ibid.*

and related devices; research and development of high efficiency low cost hydrogen energy processes; and research and development of technologies for the use of hydrogen application in energy storage systems.

There is also a need to ensure that there is a balanced renewable energy portfolio so as to achieve the positive development impacts in terms of job creation and avoided imports.[96]

A number of administrative challenges have still to be overcome. Renewable Energy planning needs to be integrated into the overall energy planning process. At the moment there are a number of plans such as the Power Development Plan 2010–2030 and the Alternative Energy Development Plan 2012–2021. There also appears to be duplication in the activities between the various agencies within the Ministry of Energy. At the moment many of the processes within these agencies lack transparency.

There is also a need for Thailand to find a way to manage the costs of its incentives without making what was formerly a simple support scheme unpredictable.[97]

## [C] Solutions

The World Economic Forum has undertaken a recent study of the energy industry in Thailand and recommended A New Energy Architecture.[98] The first recommendation was that there is a need to build in flexibility to the 10-year Alternative Energy Development Plan to take account of the rapid changes that are taking place in the renewable energy sector.[99] They considered that Small Power Providers (SPPs) and Very Small Power Providers (VSPPs), should be more closely regulated and alternative financing support mechanisms such as the reverse auction Feed-in Tariff as used in India, should be considered for new energy technologies.[100]

Interconnection to the grid must be smooth as grid integration of a large share of intermittent renewable energy sources calls for: a sound policy and regulatory framework that provides interconnection standards and financial incentives to the grid companies; coordinated generation-transmission planning; and technology solutions such as smart grids, energy storage, pump storage and grid-friendly wind turbines with better power factor control and grid fault management capability to reduce disturbances to the grid.[101]

As Thailand has committed to building nuclear power, the report recommends that there should be a focus on capacity-building to lay the foundations of the nuclear sector with a focus on all areas of the nuclear energy supply chain.[102]

---

96. Weischer, *supra* n. 2, at 7.
97. *Ibid.*, 4–5.
98. World Economic Forum New Energy Architecture: Thailand, October 2012, http://www.weforum.org/reports/new-energy-architecture-thailand (accessed 30 Jun. 2013).
99. *Ibid.* 34.
100. *Ibid.* 36.
101. *Ibid.* 38.
102. *Ibid.* 44.

Finally, the report recommends the need to foster understanding about energy issues targets have to be translated into a language that consumers understand.[103] Demonstrate and engender role model change through pilot programs that bring local benefits.[104]

It is clear that as well as the essentially technical solutions as recommended by the World Economic Forum Report, there is also a need for regulatory and administrative reform. Whilst the Ministry of Energy is the regulatory authority, there are a number of subordinate authorities that are involved in developing policy. This has led to fragmentation and the issue of development plans such as Power Development Plans and Alternative Energy Plans that do not appear to be completely coordinated. When unforeseen issues arise, additional administrative hurdles are introduced.[105] On the face of it, this is not necessarily an unreasonable response. The problem is that the approval process has become somewhat opaque, and the integrity of the process can be jeopardized.

After nearly three years, the detail of the new Feed-in Tariff regime has still not been released. Detailed information on the scheme should be made more readily available, and the information that is should include full technical papers and not just PowerPoint presentations, of which there are plenty.

As much of the current technology is imported Thailand requires a comprehensive program of institutional strengthening to develop the local renewable energy industries as well as the staff to operate both current and future facilities.

## §6.04 CONCLUSION

Thailand has a rapidly developing renewable energy sector supported by Government policy which sees renewable energy as salable commodity.

It is not sufficient to have a vision. It must be supported by a transparent and efficient regulatory and administrative framework as well as have innovative financial mechanisms. Initial issues with the large initial response for the offer to the supply of solar energy followed by the slow response to actually build and operate the generating facilities have led to increase administrative imposts and a slow response to develop streamlines approval processes and release of details of the new Feed-in Tariff mechanism.

---

103. *Ibid.*, 58.
104. *Ibid.*, 60.
105. Tongsopit & Greacen, *supra* n. 4.

CHAPTER 7
# Feed-In Tariff for Indonesia's Renewable Electricity

*Madjedi Hasan & Anton S. Wahjosoedibjo*

## §7.01 INTRODUCTION

Two of the issues facing Indonesia in its economic development today are an acute electricity shortage and global warming and climate change. Growing concerns about climate change and awareness about environmental problems led to the implementation of actions favorable to the new and renewable energies (long-lasting energies) for reduction of the greenhouse gas emissions in particular. Subsequently, electricity market in Indonesia is expected to move from a monopoly fossil-fuel generation base to a more competitive structure with an increasing share of renewable energy. The increase in the utilization of renewable energies is also motivated by the depleted oil reserves, which in the past have been the main source of finance and energy source for the country's development. Also, Indonesia has many forms of Renewable Energy (RE), such as: water/hydro, solar, wind, biomass, geothermal, and ocean waves.

In addition to hydro power, the abundant RE resource in Indonesia is the geothermal. The country is home to around 265 volcanoes and is supposed to possess more than 40% of the world's geothermal power. Currently identified resources could provide about 30,000 MW of power in Indonesia, and undiscovered resources might provide larger than that amount. Of these, presently, the country can only generate about 1,200 MW (4.6%) power. For solar, the technology is maturing but requires high capital expenditure of up to USD 5,000/kW installed capacity and faces land indemnity complication in municipality areas. This high cost makes solar cell development not viable in general conditions. However, RE for remote areas may use availability and scalability of energy as the main criteria instead of energy cost. The solar power is best used in direct sun light, remote areas.

As a tropical country, Indonesia has the most abundant energy for photosynthesis on the same surface area. Biomass energy can be in the form of bio-gas, bio-liquid, and bio-solid fuels. However, the policy makers in Indonesia are still thinking for the biomass in the direction of energy from waste instead of crop to energy. This judgment has made biomass low priority and thus not properly employed. In addition, Indonesia still applies subsidy on petrol, which made bio-ethanol less preferred.

This report discusses the implementation of Feed-in Tariff (FIT) for renewable energy in Indonesia, its current status, and challenges. As policy mechanism designed to accelerate investment in renewable energy technologies, FIT is a guarantee that renewable energy producers will be able to sell the electricity they generate at a price set in advance by the government under the long-term contract. As part of an energy policy that focuses on supporting the development and dissemination of renewable power generation, the FIT scheme providers of energy from renewable sources, such as solar, wind, or water, receive a price for what they produce based on the generation costs. This purchase guarantee is offered generally on a long-term basis, ranging from 5 to 20 years, but most commonly spanning 15–20 years.

## §7.02 ENERGY DIVERSIFICATION

At present, Indonesia's installed power capacity totals 43,528 MW, generated by Perusahaan Listrik Negara (PLN) or State Owned Electricity Company (31,943 MW), Independent Power Producer or IPP (9,856 MW), and others (1,729 MW). The source of fuel consists of 50.7% coal, 22.2% gas, 15.8% fuel oil, 6.4% hydro-electric, 4.7% geothermal, and 0.1% other energy. The high fuel oil consumption is due to the use of diesel engines that carry based load, especially outside the Java-Bali system, and liquid fuel due to unavailability of gas in the Java-Bali system.

In order to reduce the dependency on oil-based fuels for electricity generation, the Government of Indonesia (GOI) has issued a number of *Presidential Regulation (PR)* addressing the energy diversification. The first was PR Nr. 5/2006, which set the following targets by the year 2025: energy elasticity of less than one; optimal energy mix by reducing oil to 20% while increasing gas (30%), coal (33%), liquefied coal (2%), bio-fuel (5%), geothermal (5%), and new and other renewable energy such as biomass, nuclear, hydro-electric, solar, and wind (5%). The government will take measures to add the capacity of micro hydro power plants to 2,846 MW by 2025, biomass of 180 MW by 2020, wind power of 0.97 GW by 2025, solar power of 0.87 GW by 2024, and nuclear power of 4.2 GW by 2024. The total investment needed for this development of new and renewable energy sources up to the year 2025 is projected at USD 13,197 million.

In 2006 and 2010 the GOI issued two *Fast Track Programs (FTP)*, involving accelerated construction and installation of new power plants, each with a total capacity of 10,000 MW to reduce oil fuel consumption. The First FTP consists of 30 coal fired power plants. The Second FTP includes construction and installation of 92 new power plants, of which 51 are geothermal power plants and six hydro-electric power plants. The projects were listed in the Minister of Energy and Mineral Resources

(MEMR) Regulation Nr. 2/2010, which has then been replaced two times (the latest being MEMR Regulation Nr. 1/2012, effective through 31 December 2014) and has a target to reduce $CO_2$ emission.

After the Fast Track Program, the GOI expects a 56% increase in overall energy investment by 2014. Investments will likely leverage private investment with public money to focus on increasing energy reliability and to meet the government's 90% electrification target by 2020. To further international investment in the renewable energy sector, the government has also announced a tariff cap of USD 0.097/kWh for geothermal energy. In late 2010, the GOI declared *vision 25–25*, whereby by the year 2025, the new and renewable energy contribution to the energy mix will reach 25% of the total energy used, as compared to 17% in the *2005 National Energy Policy*. Table 7.1 shows the renewable energy contribution to the energy mix through 2025.

Table 7.1  Projected Renewable Energy Contribution in Energy Mix through 2025

| Fuel Type | Unit | 2007 | 2010 | 2025 | Potential |
|---|---|---|---|---|---|
| Biodiesel | Kilo Liter | 133 | 482,000 | 16,371,559 | |
| Bio-ethanol | Kilo Liter | – | 296,000 | 6,876,005 | |
| Other Biomass | MW | 445 | 500 | 870 | 49,810 |
| Geothermal | MW | 1,052 | 1,260 | 12,332 | 28,500 |
| Wind | MW | 2 | 4 | 256 | 9,290 |
| Solar | MW | – | 12 | 250 | 4.8 kWh/m3/d |
| Small Hydro | MW | 210 | 245 | 2,486 | 770 |
| Large Hydro | MW | 4,200 | 4,380 | 14,516 | 76,170 |

*Source*: DJEBTKE Vision 25–25.

As Table 7.1 shows, the Nuclear Power Plant (NPP) is not included in the energy mix through 2025, despite the fact that the activities towards the peaceful use of nuclear energy in the country dated back to the early 1960s, when Indonesia began to operate the first research reactor. Also, with increasing competition for oil and gas and the negative impact of carbon pollution, nuclear power is still considered by the GOI as a matter of survival, both in terms of growing energy demands and environmental security.

With growing electricity shortages, Indonesia was seen as the nuclear powerhouse of the future. In order to meet the increasing electricity demand, Indonesia has planned for four NPPs by 2024. But, local resistance to NPP has emerged since, which led to the cancellation of the plan to construct an NPP in Muria (Central Java). Like in many Asian countries, while there has been increased local and environmentalist opposition to nuclear powers, it has not been sufficient to reshape government policies that promote nuclear power. The GOI still hopes to be able to relocate the site to another place; however, the events in Fukushima have led this to a still stand – at least contemporarily.

## §7.03 FIT SCHEME

The main goal of a FIT program is to clarify and minimize risk factors, which will ultimately increase investment security and assist the project with securing lowest cost capital. As a general rule, lower risk translates into a lower required return for investors and a lower cost of energy from a project. Characteristics that will have the biggest impact on financing are grouped into three sets of issues, namely FIT structure, contract terms, and conditions. The available tariff may change over time. Therefore, successful FITs will encourage rapid, sustained, and widespread RE development.

FIT policies typically include three key provisions: (1) guaranteed access to the grid; (2) stable, long-term purchase agreements (typically, about 15–20 years); and (3) payment levels based on the costs of RE generation.[1] In some countries such as Germany, they may also include streamlined administrative procedures that can help shorten lead times, reduce bureaucratic overhead, minimize project costs, and accelerate the pace of RE deployment.[2]

There are four main approaches used to set the overall FIT payment to RE developers, and these are as follows:

(1) Based on the actual *levelized cost of RE generation*, plus a targeted return, which is set by the regulators or policy makers. The advantage of this approach is that the FIT payments can be specifically designed to ensure that project investors obtain a reasonable rate of return, while creating conditions more conducive to market growth. This would result in quick and substantial RE capacity expansion at both distributed and utility-scale levels. This approach is the most commonly used in the EU, and has been most successful at driving RE development around the world.[3]

(2) Based on estimated *value of the renewable energy generation* either to society or to the utility. This value can be defined in a number of ways, either according to the utility's avoided costs, or by attempting to internalize the 'externality' costs of conventional generation. Also, externality costs can include things such as the value of climate mitigation, health and air quality impacts, and/or effects on the energy security. The value-based approaches may not match the actual RE generation costs, and may provide insufficient payments to stimulate rapid market growth. Alternatively, they may provide payments that are higher than generation costs, leading to cost-inefficiency.

---

1. Mendonça (2007), *Feed-in Tariffs: Accelerating the Deployment of Renewable Energy*. EarthScan, London.
2. Fell, H.J. (2009), *Feed-in tariff for renewable energies: an effective stimulus package without new public borrowing*. Accessed 19 Jun. 2009, at http://globalwarmingisreal.com/ EEG_Fell_09.pdf.
3. Klein, A. (2008). *Feed-in Tariff Designs: Options to Support Electricity Generation from Renewable Energy Sources*. Saarbrucken, Germany: VDM Verlage De. Muller Aktiengesellschaft & Co. KG.

(3) Setting FIT payments as a simple, *fixed-price incentive* that offers a purchase price for renewable electricity that is based neither on generation costs, nor on the notion of value.[4]

(4) Based on the results of a*uction or bidding process*, which can be applied and differentiated based on different technologies, project sizes, etc. and is a variant on the cost-based approach. Both India and China are experimenting with this approach.[5]

As can be expected, the most effective FIT policies to meeting the objectives are those designed to cover the RE project cost, plus an estimated profit (item 1).[6] This effectiveness arises from the fact that developers will not invest unless they are relatively certain that the revenue streams generated from overall electricity sales are adequate to cover costs and ensure a return.[7] To the extent that FITs provide a single long-term revenue stream, they can reduce investment risks and increase the rate of RE deployment.

However, other procurement mechanisms can theoretically achieve the same objective, provided they offer the same, or similar, foundational elements. In the case the objective is to limit policy costs, this could be accomplished by establishing payment levels targeting only the most cost-effective technologies, or limit deployment to areas with the best combination of attributes, including resource, proximity to transmission, etc. FIT payments can also be differentiated by *technology type, project size, resource type, and quality*.[8]

Many FIT implementation options center on the utility's role. Several countries use a FIT policy *purchase obligation* that requires utilities (such as in Indonesia) to purchase the entire output from eligible projects, which guarantees eligible project owners that they will be able to interconnect their projects to the grid. Caps may be imposed either on the total capacity of RE allowed (usually differentiated by technology type), on the maximum individual project size (also often differentiated by technology type), or according to the total program cost (either total dollars per year, or for the multiyear program).

Also, guarantees can also be given to transmission system operators (TSOs) to offer *guaranteed grid connection*. In this respect, it is important to provide clear protocols surrounding *transmission and interconnection issues*, which ensure that RE

---

4. Cory, K.; Couture, T.; Kreycik (2009), C. *Feed-in Tariff Policy: Design, Implementation, and RPS Policy Interactions*. Golden, CO. National Renewable Energy Laboratory Technical Report No. TP-6A2-45549, http://www.nrel.gov/docs/fy09osti/45549.pdf.
5. Kann, S. (2010). *The India Solar Market: How Big and How Soon?* GreenTech Media Web site. http://www.greentechmedia.com/articles/read/the-india-solar-market-how-big-and-how-soon/.
6. Klein, A. (2008). *Feed-in Tariff Designs: Options to Support Electricity Generation from Renewable Energy Sources*. Saarbrucken, Germany: VDM Verlage De. Muller Aktiengesellschaft & Co. KG.
7. Deutsche Bank. (2009). *Paying for Renewable Energy: TLC at the right price*, DB Climate Change Advisers (DBCCA). Frankfurt, Germany; December 2009. Accessed http://www.marsdd.com/default/dms/ events/paying_for_renewable_energy.pdf.
8. Mendonça (2007), *Feed-in Tariffs: Accelerating the Deployment of Renewable Energy*. EarthScan, London.

projects can be connected to the grid in a timely way that minimizes bureaucratic overhead and fosters more efficient project sitting.

## §7.04 OVERVIEW OF FIT POLICIES IN INDONESIA

In Indonesia, the FIT program was first introduced in 2002, when the MEMR issued a regulation for small scale renewable power plant below or equal to 1 MW. The FIT was used in the 12 Power Purchase Agreements (PPA) with total installed capacity around 5.5 MW, all from hydro power plants. The tariff was set up as a base on the electricity basic production cost (Harga Pokok Produksi or HPP) of PLN, and therefore every region of PLN has different HPP. The tariff was set at 60% of the HPP in the low voltage interconnection and 80% of the HPP in the middle voltage interconnection.

In 2006, the capacity parameter was increased to include the medium scale renewable power plant up to 10 MW under the same tariff system with all technologies being eligible for the tariff that has a long-year maturity. But the problem with this financial incentive is not the level of support provided but the regulated end user energy prices, with a growing concern that unless the government increases these prices the FIT system will become economically unsustainable.

In March 2009, MEMR issued a new regulation that the tariff will be set up based on the PLN's own price on the principle of economic value fairness price, and on the negotiation of business. Although the tariff has been set up more attractive and fair, the issues remained with respect to how the tariff must be developed, how long it will survive, and whether or not PLN will accept it.

For example, the costs of the coal-fired generation are just over half that of renewable resources and for PLN, coal is a far more economically attractive generation source. Also, while PLN is obligated to off-take renewable generation from IPPs, the disparity between renewable generation costs and end-user prices could well force the utility to switch its own production to coal in order to reduce losses, which would ultimately undermine the government's renewable and emission objectives.

In Indonesia, the design of tariff is differentiated by the technology and resource based on generation cost. This will provide sufficiently high tariffs on the one hand, avoiding wind fall profits on the other hand. As can be seen in Tables 7.2 and 7.4, there are distinct FIT rates for geothermal and other renewable resources. Furthermore, as described in the MEMR Regulation Nr. 44/2012, the other renewable is further differentiated by biomass and bio-gas, waste and hydropower, wind, and solar. The waste is further differentiated by technology such as zero work and sanitary landfill.

The following lists the various regulations and issues dealing with the FIT system for each resource. For the purpose of discussion, these resources are divided into two groups based on the importance in meeting the country's electricity demand, namely geothermal and miscellaneous resources such as hydro power, solar, wind, biomass and municipal waste.

## [A] Geothermal

(a) MEMR Regulation Nr. 2/2011, authorizing PLN to purchase geothermal electricity at the tendered price and obligating PLN to develop geothermal Power Purchase Agreement (PPA) model.[9]

(b) Minister of Finance Regulation Nr. 3/2011 on guideline for the management and accountability of Geothermal Fund Facility (GFF). Administered by Pusat Investasi Pemerintah (PIP) or Government Investment Agency, under the scheme, GOI allocated in 2011 and 2012 State Budget a fund of USD 220 million to help mitigating risk by enhancing data of geothermal prospects (including drilling up to three pre-exploratory to reservoir depth).

(c) MEMR Regulation No. 22/2012, which amends previous regulation on Feed-In Tariff (FIT) for geothermal electricity from a nation-wide cap at USD 0.97/kWh to fixed tariff as shown in Table 7.2 below.

*Table 7.2   Feed-In Tariff for Geothermal*

| Location | Tariff (USD cents per kWh) | |
| --- | --- | --- |
|  | *High Voltage* | *Medium Voltage* |
| Sumatra | 10.0 | 11.5 |
| Java, Madura, Bali | 11.0 | 12.5 |
| South, West and South East Sulawesi | 12.0 | 13.5 |
| North and Central Sulawesi and Gorontalo | 13.0 | 14.5 |
| West and East Nusa Tenggara | 15.0 | 16.5 |
| Maluku and Papua | 17.0 | 18.5 |

(d) As Table 7.2 shows, the differentiation includes geographic differentiation in that there are different rates for system on Java and systems built in other major islands in Indonesia. The adjustment, however, is made based on the avoidance cost principle, namely the cost of oil-based fuel (diesel) for electricity generation. The differentiation is also based on the quality and size of potential resource, where the geothermal fields in Sumatra and Java have been assumed to have high enthalpy resource and larger reserves. While the geothermal resources in East Indonesia are generally low enthalpy resources associated with non-volcanic environment, thereby lower potential and higher development costs.

---

9. *See* Implementation of FIT.

(e) In the non-geothermal sources, the differentiation is made by adjustment factor 'F', which ranges from 1.0 to 1.5 for seven Indonesia's major islands. The factor 'F' reflects the additional cost associated with remoteness from the capital, however the methodology used in determination of F is not known.

(f) The new tariff regime will be applicable to new developers that are issued their geothermal business licenses (Izin Usaha Panas Bumi or IUPs) after 23 August 2012 and present developers which hold other forms of geothermal authorizations, permits or contracts issued before the 2003 Geothermal Law who wish to pursue expansion projects, extension of the PPA or who have already signed PPAs with PLN (and whether or not the plant has started producing electricity or steam), provided that the terms of the PPA provide that the parties may agree to changes in the price of electricity or steam. The new tariff is also applicable to developers that hold IUPs that will 'implement' PPAs, provided that the price amendment is agreed between the parties and the PPA makes possible the use of the new tariffs.

(g) Article 1.3 of MEMR Regulation No. 22/2012 states that the parties who are entitled for the new pricing regime include:
 - New IUP holder (to be issued after 23 August 2012).
 - Existing developers (before Promulgation of Geothermal Law Nr. 27/2003) who are planning for plant expansion.
 - Existing developers (before Promulgation of Geothermal Law Nr. 27/2003) who are nearing contract expiration and to be extended.
 - Existing developers (before Promulgation of Geothermal Law Nr. 27/2003) who have the PPA in place but have not produced steam or electricity provided there is a clause in the agreement that allows the price negotiation.
 - Existing IUP holders who are in the process of negotiation for Power Purchase Government (PPA), subject to the Parties' agreement and the price adjustment is allowed under the said PPA.

(h) The new FIT would only apply to new projects or extension contracts between IPPs and PLN. There is the possibility to revise existing pricing through negotiations. Note Article 4 of the new Regulation allows PLN to buy the power at the higher price or in excess of the amounts specified in the regulation, subject to the agreement by the parties and based on PLN's own estimates and approved by the Minister.

(i) In response to the industry's demand, the GOI is now considering to amend the MEMR Regulation No. 22/2012 on FIT. The amendment will involve differentiation of FIT payments to account for the differences in size and types or the choice of technology used. In addition to distinction between high temperature (higher than 225°C) and medium/low temperatures (lower than 225°C), the prices will also be structured in accordance with the size as shown in Table 7.3 below.

*Table 7.3    Proposed New Feed-In Tariff for Geothermal*

| Capacity | Feed-in Tariff –c$/kWh | |
|---|---|---|
| | High Temperature | Low/Medium Temperature |
| Greater than 55 MW | 10.5 | 13.5 |
| Between 20 MW–55 MW | 11.5 | 15.0 |
| Between 10 MW–20 MW | 13.5 | 16.0 |
| Smaller than 10 MW | 17.0 | 19.0 |

(j) In geothermal power development, the FIT policies have been designed to provide the guarantee of a stable revenue stream over the life of the generation plant, thereby lowers the investment. The combination of a guaranteed purchase, a pre-determined payment price, and a standardized off-take agreement can relieve some of the cost, risk, and pressure associated with overall project development since the project does not need to compete for or negotiate a contract before final project costs are known.

**[B]    Other Renewable Energies**

(a) MEMR Regulation Nr. 4/2012 provides Feed-in Tariff of Renewable Energy (other than geothermal) electricity for plant capacity below 10 MW. PLN is obliged to purchase Renewable Energy electricity generated by private and state own enterprises or cooperatives at above prices without further negotiation. The tariffs for hydro power, solar, and wind are the same as those stated in previous MEMR Regulation No. 31 Year 2009. GOI has indicated that FIT for hydro power, solar, and wind will also be increased (*see* Table 7.4).

Also, the differentiation includes geographic differentiation in that there are different rates for system on Java and systems built in other major islands in Indonesia. The differentiation is made by adjustment factor 'F', which ranges from 1.0 to 1.5 for seven Indonesia's major islands. The factor 'F' reflects the additional cost associated with remoteness from the capital, however the methodology used in determination of F is not known.

*Table 7.4    Feed-In Tariff Renewable Energy*

| RE Type | Price Rp/kWh | | F Factor | | | | | | |
|---|---|---|---|---|---|---|---|---|---|
| | MV | LV | A | B | C | D | E | F | G |
| Biomass – Bio-gas (Non Municipal Solid Waste) | 975 × F | 1,325 × F | 1.0 | 1.0 | 1.2 | 1.2 | 1.2 | 1.3 | 1.3 |
| Municipal Waste – Zero Waste Technology | 1,050 × F | 1,398 × F | 1.0 | 1.0 | 1.2 | 1.2 | 1.2 | 1.3 | 1.3 |

| RE Type | Price Rp/kWh | | F Factor | | | | | | |
|---|---|---|---|---|---|---|---|---|---|
| | MV | LV | A | B | C | D | E | F | G |
| Municipal Waste – Sanitary Landfill | 850 × F | 1,198 × F | 1.0 | 1.0 | 1.2 | 1.2 | 1.2 | 1.3 | 1.3 |
| Others: Hydro, Wind | 656 × F | 1,004 × F | 1.0 | 1.2 | 1.3 | 1.2 | 1.3 | 1.5 | 1.5 |

A: Java-Madura-Bali; B: Sumatra; C: Kalimantan; D: Sulawesi; E: Nusa Tenggara; F: Maluku; G: Papua.

(b) For solar photovoltaic power plants, on 12 June 2013 the GOI issued MEMR Regulation Nr. 17/2013 that stipulates FIT in the form of a price cap at US 25 cents/kWh. If the plants use photo voltaic module with local components of at least 40%, an incentive will be granted by increasing the maximum tariff to US 30 cents/kWh (this might violate the WTO Rules[10]). The purchase of electricity from the solar photovoltaic power plants by PLN is based on tendered Capacity Quota, which is the maximum capacity of photovoltaic power plant that can be interconnected to the PLN grid or subsystem. The winning bid price will be formulated in the Power Purchase Agreement which is valid for 20 years of operation.[11]

[C] **Duration of FIT**

Indonesia's FIT regulation is also silent with respect to the duration of FIT. Also, Indonesia's FITs do not include a schedule of 'declining payment' under which payments decline over time such that a generator that comes on line say in 2010 would get lower payment than a generator that comes on line in 2009. In this way, cost reductions due to experience curve effect are included in the policy, leading to lower burden on the electricity consumers. In addition to lower burden for the consumers, the declining tariff will be an incentive for technological improvements, transparency, and incentives to build early. The disadvantage includes that the system cannot respond to dramatic technology price changes and can lead to high administrative complexity (*see* Tables 7.3 and 7.4 for updated rate schedule in 2003).

---

10. *See* WTO's Highest Court Decision in the case Japan versus Canada (6 May 2013), stated that domestic content requirements that require some generators to source up to 60% of equipment from the Canadian province of Ontario under its feed-in-tariff (FIT) program are inconsistent with international trade rules.
11. *See also* discussion on Implementation of FIT and Other Barriers to Renewable Electricity Development.

## [D] Price Escalation

A pertinent issue related to MEMR Regulation Nr. 22/2012 is with respect to electricity price escalation. The opponent to price escalation may argue that although the operational expenses and subsequent capital costs may increase, the computation for FIT has been made based on levelized costs of energy, which is defined as the constant price per unit of energy that causes the investment to just break-even: earn a present discounted value equal to zero. In other words, present discounted value of energy produced times the levelized cost equals the present discounted value of the fixed and variable costs over the life of the investment.

On the contrary, the supporters to price escalation may argue that a large portion of the costs involved in developing a geothermal project are capital costs invested up front, thereby there may be no strong basis for these costs being the subject of indexation over time. However, all geothermal projects do involve varying degrees of operational expenses and drilling make-up wells to maintain the geothermal resource, which tend to increase. Accordingly, a flat tariff structure without indexation (over 30-year PPA terms) may expose developers to a mismatch between flat revenue and increasing operating costs.

Article 39 of the Government Regulation No. 14 Year 2012 on Electricity Supply Business Activities stipulates that the price of electricity as approved by the minister, governor, or regent/city mayor, in accordance with their respective authority, can be adjusted due to certain changes of cost elements on the basis of mutual agreement, which should be stated in the PPA. Such price adjustment can be carried out upon receiving approval from the minister, governor, or regent/mayor in accordance with their respective authority. The foregoing clause provides the legal basis for price escalation in a PPA. In order that the price escalation would not cause instability, such price escalation can only be made effective if there are significant changes of a certain economic parameter, such as changes of Consumer Price Indices (CPI) for certain commodities and Exchange Rate exceeding ± 20%.

The latter is also true as Indonesia's system of FIT as outlined in the MEMR Regulation Nr. 22/2012 has been developed on the basis of avoided costs and not based on the actual levelized cost of renewable energy generation, plus a stipulated return (set by the policy makers, regulators, or program administrators). Also, since the development costs for geothermal resources will be dependent on size and enthalpy of the resources, differentiation of FIT payments to account for the differences in size and types or the choice of technology used can help the acceleration of small capacity geothermal field and low enthalpy resources. Therefore, this may be considered not a true FIT, although the Tables 7.3 and 7.4 seem to show that the FITs have been developed based on actual cost.[12]

---

12. *See also* discussion in the 'Implementation of FIT'.

### [E] Key Input Parameters

Unlike in some countries, where the process is driven by non-governmental consultants, in Indonesia, the development process of FIT is driven by the MEMR. Although the regulations on FIT were prompted by the industry's demand and at the outset of the process MEMR office generally convenes a series of discussion with the industry through the Industry Association, the recommendations by the industry are not binding. Such approach to tariff setting has indeed some advantages, as it can move forward more quickly because there is less opportunity for stakeholder intervention.

On the other hand, the disadvantage is that the stakeholder process is not iterative and does not have a chance to comment on the cost model used. This has been experienced in Indonesia, given the fact that the GOI is now preparing a new MEMR on FIT for geothermal, to replace the current MEMR Regulation 22/2012, as the said regulation still leaves uncertainties.

## §7.05 IMPLEMENTATION OF FIT

As stipulated in the Government Regulation, any procurement for goods, services, or works from external sources will involve the bidding process known as 'tendering'. It is favorable that the goods, services, or works are appropriate and that they are procured at the best possible cost to meet the needs of the purchaser in terms of quality and quantity, time, and location. Corporations and public bodies often define processes intended to promote fair and open competition for their business while minimizing exposure to fraud and collusion.

Like other countries, in Indonesia, strict rules on procurement must be followed by public bodies, with contract value thresholds determining the processes required (relating to advertising the contract, the actual process, etc.). Given the new FIT regime, it appears that in the geothermal activity the potential developers will no longer be required to bid the tariff that they are willing to accept, since the applicable tariff has been set up based on location and connecting transmission as set out in Table 7.2.

This would raise a question as to what would be the criteria to be used in determining the winner of a tender? Under such circumstances, new bid evaluation criteria must be established, which may include work program and expenditures commitment and signature bonus, and technical and financial capability. Given experiences in oil and gas, new bidding criteria will include *the work program and expenditures commitment and bonus payment*. For effectiveness and from a transparency perspective, such commitment shall be supported by appropriate bank security in the form of bank guarantees or escrow accounts.

Furthermore, confronted with the Government Regulation No. 59/2007 that stipulates that the bid winner is determined by the lowest bid price (Government Regulations has higher hierarchy than Minister Regulations), the new FIT schemes will in its application be in the form of a cap price instead of a fix price. The new regulation of electricity tariff for solar photovoltaic reflects this condition. The draft new Geothermal FIT as shown in Table 7.3 will also be considered as a capped price.

## §7.06 OTHER BARRIERS TO RE DEVELOPMENT

Although Indonesia's renewable energy resources are substantial and the government has committed to increasing the deployment of renewable energy by including among others the introduction of FIT scheme for renewable electricity, several barriers must still be overcome for the country to reach its full potential. Unlike fossil-fuel projects, most RE projects require integrated efforts and risk management between upstream and downstream activities. Some may involve basic development issues as will be explained below.

The installation of power plants will involve various stages of work and clearances which we need to go through including project identification, project allotment, pre-feasibility report, detailed project report, techno-economic clearance, environmental and other clearances, financial closure, finalizing contracts for civil and mechanical work, and commencement of construction. Out of many steps and clearances required to set up a power project, making land available is a herculean task. It poses a major challenge, since there are many stakeholders with varied forms of interest in making this possible. Land is of basic importance and necessity when it comes to pre-requisites of power generation. Most of the projects get delayed or canceled due to the effect of non availability of land at market prices.

In addition, the government itself is sometimes a barrier to renewable energy deployment. It often struggles to implement policy initiatives, which causes bottlenecks for innovative technologies like renewable energy. Following the promulgation of autonomy laws in the late 1990s, local governments have been given a larger authority, including administering tenders for projects, but they often lack the capacity to do so transparently. Several companies have complained that tenders are worded poorly and amended after the tender is announced. The GOI is now assisting the local governments to promote better tendering practices for the government's related projects.

Moreover, one of the weaknesses in RE transactions is the requirement to have lowest electricity tariff during the tender. As a result, the bidders might not take a cautious approach to risk, thus tending to substantially lower the offered price of electricity below that which might in the event be actually necessary (as would eventually be established by the exploration activities). In the case of geothermal power, this is a direct consequence of not having executed exploration works in advance of Work Area tendering. The end result is that the party that wins the bid would have difficulties in securing the finance for the project. Of the 33 geothermal projects involving new geothermal work areas under the Second 10,000 MW fast-track power plant projects, 12 projects failed their commitment to raise a fund of USD 20–30 million per project to carry out pre-exploration work.

Other factors relate to the facts that developments of various energy sources are found in various laws and regulations, thereby they have potential for conflict in the execution. Also, land rights and ownership have been a vexing matter since Indonesia's independence, attributable to among others by the rich diversity of the local population adhering to different sets of customary rules and high-handed bureaucratic intervention. Other crucial issues include overlapping of rights on mining operating areas on one side with the rights of forestry, people's land, and others. The overlapping lands and permits and conflicting interests have ultimately delayed the activities for development of oil and gas resources, mineral, coal, and geothermal.

## §7.07 CONCLUSION

The GOI has made a breakthrough in the development of RE such as with the FIT regulation to develop geothermal and other renewable resources. However, inconsistencies in the law and regulation and heavy subsidies for energy consumption have forestalled the development of a cleaner energy infrastructure. Designing a policy structure with short-term availability and uncertainty adds to the overall risk of a project and will create a barrier entry for many developers and potential capital providers. Indonesia needs to review the structure of renewable transactions, while the GOI also provides assistance in securing the land, in particular the project that has been included in the Accelerated Power Plant Project as listed in the Presidential Decree Nr. 4 of 2010 and MEMR Regulation No. 2/2012.

# Part III Cross Country Analysis of Renewable Electricity Promotion Regime

CHAPTER 8
# Evaluation of Eleven Implemented Policy Mixtures in the Black Sea and Caspian Sea Regions for the Use of RES

*Popi Konidari*

The use of renewable energy sources is considered as an important component of the energy and climate change policy of a country. This paper examines the currently implemented policy mixtures for the deployment of RES in 11 emerging economies (Albania, Armenia, Azerbaijan, Bulgaria, Estonia, Moldova, Romania, Russia, Serbia, Turkey, and Ukraine) with the use of two research tools, the LEAP model and the multi-criteria evaluation method AMS. The AMS outcomes will show which one of these 11 policy mixtures is more effective in reaching the set national RES target. Strengths and weaknesses of each policy mixture are analyzed and conclusions focus on the common elements of the most successful policy mixtures.

## §8.01 INTRODUCTION

Energy efficiency (EE) and Renewable Energy Sources (RES) have an important role in achieving climate change mitigation targets (Umweltbundesamt, 2009). The exploitation of RES is not only a climate change policy objective, but an energy policy one also. RES are essential not only in mitigating GHG emissions, but also in advancing sustainable development by improving the access of millions to energy, reducing the dependence on fuel imported, decoupling energy costs from oil prices, decarbonizing the energy sector and contributing to energy security (Eurostat, 2013; Greenpeace, 2012; EC, 2011; CEC, 2008).

In 2009, RES accounted for 13.5% of the world's primary energy demand with 19.3% contribution in electricity generation and 25% approximately in heat supply (Greenpeace, 2012). Fossil fuels had a share of 81% of the primary energy supply and the remaining 5.5% came from nuclear energy (Greenpeace, 2012). Their contribution for the satisfaction of future energy needs under a sustainable approach is considered to increase in the majority of global or regional energy and climate change policy scenarios (IEA, 2013, 2012; SEI, 2012a; BP, 2012; Greenpeace, 2012; EREC, 2011; UNEP, 2011). Estimations of these scenarios provide a considerable contribution of RES in the global primary energy supply ranging from 23% to 31% by 2030 and from 35% to 56% by 2050 (Umweltbundesamt, 2009).

This increased future contribution is justified by the increased current trend for investments in RES projects by the governments mainly due to: (i) the potential of RES projects to create jobs and (ii) technological advances that improve competitiveness (UNEP, 2011). Global investment in renewable power and fuels increased by 17% in 2011 with developing economies having a 35% share of this investment compared to the 65% of the developed ones (Frankfurt School – UNEP Collaborating Centre for Climate & Sustainable Energy Finance, 2012). Most of the emerging economies following the example of developed countries have adopted RES targets and set up a policy mixture for the exploitation of their RES potential.

This paper examines the currently implemented policy mixtures in 11 emerging economies: Albania, Armenia, Azerbaijan, Bulgaria, Estonia, Moldova, Romania, Russia, Serbia, Turkey, and Ukraine. These countries were selected for this analysis as this paper is based on the work executed by the FP7 Funded Programme 'PROMITHEAS-4'. PROMITHEAS-4 concerns the development and evaluation of Mitigation/Adaptation policy portfolios for 12 countries (the aforementioned ones plus Kazakhstan).

The first session of this paper presents the synthesis of the policy mixtures oriented towards RES through objectives, policy instruments, and pertinent implementation network. The second session presents the LEAP model, the basic key assumptions that were adopted for estimating the use of RES in year 2020 as a result of the currently implemented policy mixture. The third session refers to the multi-criteria evaluation method AMS under which all policy mixtures are evaluated. The fourth session concerns the evaluation of these policy mixtures using the two research tools. Finally, the outcome of this evaluation is discussed, analyzed, and leads to: (i) the most effective policy mixture out of the 11 ones and (ii) those common elements that configure the most effective policy mixtures.

*Figure 8.1  Map of the 11 Emerging Economies*

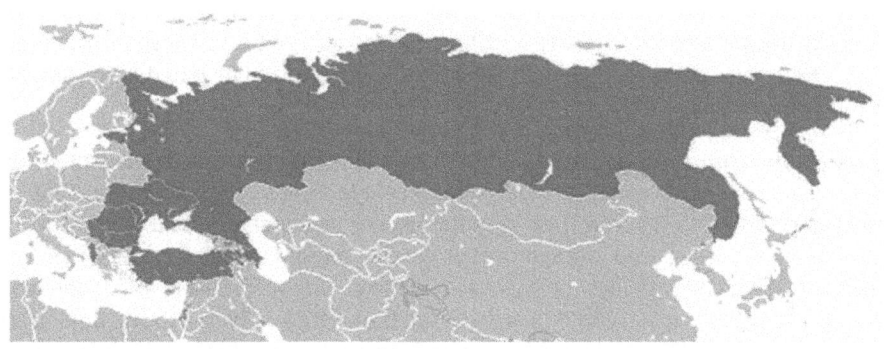

(http://upload.wikimedia.org/wikipedia/commons/c/c3/BlankMap-World.png).

## §8.02  POLICY MIXTURES FOR THE USE OF RES

### [A]  Objectives

Four of these countries (Albania, Moldova, Serbia, and Ukraine) are members of the Energy Community (EnC), two are observers to the EnC[1] (Armenia, Turkey) and three are EU Member States (Bulgaria, Estonia, and Romania). The first four have determined their RES targets by calculating them with the EU methodology and reflecting an equal level of ambition.[2] During the 10th Ministerial Council for the EnC members, which was held on 18 October 2012, the EU RES Directive 2009/28/EC was adopted and national RES 2020 targets for the nine Contracting Parties of the EnC were set (Decision 2012/04/MC-EnC on the implementation of Directive 2009/28/EC and amending Article 20 of the EnC Treaty[3]).

For the observer countries the RES targets according to the Directive were determined similarly, but there are no such official commitments yet. All official commitments are presented in Table 8.1 along with those of the remaining two countries that are neither EU Members States nor EnC Parties (Azerbaijan and Russia).

For the better understanding of these objectives (Table 8.1) the following definitions provided by the Directive 2009/28/EC are quoted:

- *'energy from renewable sources'* means energy from renewable non-fossil sources, namely wind, solar, aerothermal, geothermal, hydrothermal and ocean energy, hydropower, biomass, landfill gas, sewage treatment plant gas, and biogases; and

---

1. http://www.energy-community.org/portal/page/portal/ENC_HOME/MEMBERS.
2. http://www.energy-community.org/portal/page/portal/ENC_HOME/AREAS_OF_WORK/RENEWABLES/RES_Implementation.
3. http://www.energy-community.org/pls/portal/docs/1766219.PDF.

- 'Gross Final Consumption of Energy (GFCE)' means the energy commodities delivered for energy purposes to industry, transport, households, services including public services, agriculture, forestry and fisheries, including the consumption of electricity and heat by the energy branch for electricity and heat production and including losses of electricity and heat in distribution and transmission.

The following terms need also to be defined for understanding the different expressions that the countries have used for their objectives. *Final Energy Consumption* (FEC) is defined (according to Eurostat[4]) as the total energy consumed by end users, such as households, industry, and agriculture. It is the energy which reaches the final consumer's door and excludes the part used by the energy sector itself. When the word 'gross' is used also (i.e., GFCE) then apart from the energy delivered to the final users, the energy consumed by the energy sector and energy lost due to the transmission and distribution are also included.

The terms 'Total energy consumption', 'Total primary energy supply', and 'gross inland energy consumption' are equivalent and can be used alternately. *Total energy consumption*[5] represents the quantity of all energy necessary to satisfy inland consumption and is the result of adding production and imports, but minus exports, and international marine bunkers and plus/minus stock changes.

### [B] Policy Instruments and Implementation Network

Each country has a different policy mixture to support the use of RES. The whole set of policy instruments that these 11 countries use – along with the relative rules and influencing mechanisms – includes (Table 8.2): (i) Financial policy instruments (Feed-in-tariffs, premium schemes, capacity based schemes, tradable green certificates, ERUs, CERs, tax reductions/exemptions, investment aid); (ii) Regulatory policy instruments (Certificate/Guarantee of origin, privileged producers, Power Purchase Agreement (PPA), minimum share of national sources, certificate, Quota obligation,[6] obligatory purchase). For the monitoring, supervision, and implementation of these policy instruments each country has established its set of pertinent authorities.

---

4. http://epp.eurostat.ec.europa.eu/statistics_explained/index.php/Glossary:Final_energy_consumption.
5. http://www.eea.europa.eu/data-and-maps/indicators/total-energy-consumption-outlook-from-iea.
6. Governments impose an obligation on consumers, suppliers or producers to source a certain percentage of their electricity from RES (RES-E). This obligation is usually facilitated by tradable green certificates. RES-E producers sell the electricity at the market price, but can also sell green certificates, which prove the renewable source of the electricity. Suppliers prove that they reach their obligation by buying these green certificates, or they pay a penalty to the government (CEC, 2008).

## [C] Albania

The national policy mixture includes:

- *Exemptions from custom taxes* for machinery and equipment used for construction of RES with installed power of not less than 5 MW (Law No. 8987, 24 December 2002 (OJ 90, page 2645)).
- *Feed-in-tariffs (FITs)* (Government Decree no. 27, 19 January 2007).
- *Standard long-term PPA:* It concerns all new Small Hydro Power Plants (SHPP) with installed capacity less than 10 MW for a period of 15 years and a FIT defined by methodology (Government Decree No. 27, 19 January 2007).
- *Obligatory production:* All electric power producers, with installed capacity higher than 100 MW, are obliged to produce electric power from RES (RES-E) for the electro energetic system and not less than 2% of their total electric power production from other sources (Law no. 9072, 22 May 2003 (OJ 53, page 2120)).
- *Exemption from the excise tax:* It concerns bio-fuels used in transport till 2018, while no custom duties and VAT shall be applied for equipment and machineries used for bio-fuel production plants, equipments, and materials used by farmers for production of crops for bio-fuel production (Law no. 9876, 14 February 2008, (OJ 27, page 1281).

The Implementation Network (IN) entities responsible for RES issues are: Ministry of Economy, Trade and Energy;[7] Ministry of Environment, Forests and Water Administration;[8] Natural Agency of Natural Resources;[9] Albanian Energy Regulator;[10] Albanian Power Corporation.[11]

## [D] Armenia

The set of policy instruments consists of:

- *FITs:* They were set in force for 15 years starting from date of issue of the electrical energy production license (Energy Law (No. 148, 21 March 2001); Decision No. 52, 2 September 2001; Decisions No. 166N and No. 167N, 08 November 2005) (UNDP, 2010).
- *PPA and guaranteed access to the grid:* RES-E by licensed entities is subject to 100% purchase for a 15 years period starting from the operation beginning. The PPA between a RES generator and the Electric Networks of Armenia CJSC is signed after the construction of the power plant and obtaining the Operation

---

7. http://www.mete.gov.al/.
8. http://www.moe.gov.al/en/.
9. http://www.akbn.gov.al/index.php?&lng=en.
10. http://www.ere.gov.al/index.php?lang=2.
11. http://www.kesh.al/content.aspx?id=18.

License from the Public Services Regulating Commission (PSRC) (IFC, 2012) (same laws and decisions as above).

The IN for RES includes: Ministry of Energy and Natural Resources;[12] Ministry of Nature Protection;[13] Armenia Renewable Resources and Energy Efficiency Fund;[14] PSRC.

### [E]   Azerbaijan

The Azeri policy instruments are:

- *Exemptions from custom duties and VAT* regarding imported wind appliances and their parts (Decree No. 187, of the Cabinet of Ministers of the Azerbaijan Republic, 15 October 2005).
- *FITs* (Decree of the President of the Republic of Azerbaijan, 26 December 2005, No. 341,[15] Resolution of the Cabinet of Ministers, 30 December 2005, No. 247[16]).

The IN entities are: Ministry of Ecology and Natural Resources;[17] Ministry of Industry and Energy; State Agency for Alternative and Renewable Energy Sources.[18]

### [F]   Bulgaria

The policy mixture for RES includes:

- *Obligatory purchase:* Electricity transmission and distribution companies are obliged to purchase all RES-E and electricity from CHP. Companies trading with transport fuels are obliged to sell conventional fuels mixed with bio-fuels (Energy Act, SG 107/9.12.2007, last amendment SG 54/17 July 2012; Renewable and Alternative Energy and Biofuels Act (RAEBA), SG 49/19 June 2007, amended SG 102/22 December 2009).
- *PPAs:* The obligatory purchase of energy from RES is under PPAs with a term of: 25 years for geothermal and solar energy; 15 years for hydropower plants with capacity of up to 10 MW and other RES.[19]
- *FITs:* Electricity produced by CHP units must be purchased at preferential prices, set by the State Energy and Water Regulatory Commission (SEWRC)

---

12. http://www.minenergy.am/.
13. http://www.mnp.am/?p=80.
14. http://r2e2.am/en/.
15. www.tariffcouncil.gov.az.
16. www.tariffcouncil.gov.az.
17. http://www.eco.gov.az/en/.
18. http://abemda.az/?lang=3&id=310.
19. http://www.investnet.bg/Libraries/testLibrary/Deloitte-Bulgaria-Investment-Incentives-5.sflb.ashx.

and determined according to individual production costs plus addition. Preferential prices for RES-E have two components: at least 80% of the average end-use electricity price for the previous calendar year and a supplement set by SEWRC. The supplement is subject to change annually, but it cannot be lower than 95% of its level in previous year (Ordinance SG 17/2 March 2004, amended SG 62/31 July 2007; RAEBA).
- *Certificate of origin:* RES-E is recognized by this certificate, issued by SEWRC. It certifies the producer, amount of generated electricity, period of production, production plant and its capacity (Ordinance SG 10/6 February 2009, amended SG 85/29 October 2010; Ordinance SG 41/22 May 2007, amended SG 85/29 October 2010).
- *Priority access to the grid:* Electricity transmission and distribution companies are obliged to expand their capacity to integrate RES installations and to connect them to the grid by priority. Connection costs are paid by the RES-e producer (RAEBA, replaced by Energy from Renewable Sources Act, SG 35/3 May 2011, amended SG 54/17 July 2012).
- *Green Investment Scheme (GIS):* The Bulgarian Government can sell Assign Amount Units (AAU) and use the income from them for increasing the use of RES (specifically biomass utilization) technologies (Environment Protection Act, SG 91/25 September 2002, amended SG 53/13 July 2012).
- *Quotas (combined with existing FITs)* (Energy from Renewable Sources Act, SG 35/3 May 2011, amended SG 54/17 July 2012 – replacing the RAEBA).

The SEWRC[20] regulates mainly and exclusively RES issues.

### [G]    Estonia

The RES policy mixture includes:

- *Obligatory purchase:* The network operators could purchase in a trading period all the electricity generated by a producer of renewable energy to the extent of the operator's network losses (AgriPolicy, 2009). In 2007, a new aid scheme was introduced allowing producers of renewable energy to use the purchase obligation as before, but for selling the produced electricity and for receiving aid for the electricity sent to the grid and sold. The mandatory purchase price for RES-E rose by 42% and the possibility of using the purchase obligation was no longer restricted to the network losses (EC, 2006). The aid lasted 12 years from the start of production (EBRD, 2009a). The amendment of the Electricity Market Act in 27 February 2010 abolished the purchase obligation and only a premium of EUR 53,7/MWh is now available for RES-E. (Electricity Market Act[21] (RT I 2003, 25, 153, date: 1 July 2003; amended by RT I 2009, 39, 262, date: 24 July 2009)).

---

20. www.dker.bg.
21. http://www.konkurentsiamet.ee/?id = 19475.

- *Operating support for constructing fossil-fuel-fired CHP plants:* The support is ensured to plants that generate electricity in efficient cogeneration regime from waste as defined in the Waste Act, from peat or from the retort gas from oil shale processing. Plants that use other sources of energy receive support if their electrical capacity does not exceed 10 MW (same Act as above).
- *Obligation* of 5%-7% bio-fuel share in liquid motor fuels and of 50% bio-fuel share in liquid fuels for public transport (Liquid Fuel Act[22] (RT1 I 2003, 21, 127/1 July 2003, amended by RT I 2003, 88, 591/01 January 2004)).
- *GIS:* GIS projects concern the use of renewable energy at small boiler houses and increased share of RES-E (Regulation No. 120 of Ministry of the Environment, 22 September 2004, Appendix to the State Gazette – 2006, 33, 591; Regulation No. 115 of Ministry of the Environment, 7 September 2004 – Appendix to the State Gazette – 2004, 122, 1894; 2006, 33, 592).

The entities of the IN for RES are: Ministry of the Environment;[23] Ministry of Economic Affairs and Communications;[24] Elering.[25]

### [H]   Moldova

The RES policy mixture includes:

- *FITs:* Charges are set for RES and approved on annual basis, depending on type and capacity of production facilities, production volumes, expected delivery period and delivery of renewable energy. Tariff prices on the international market will be taken into account also when applicable (Law on Renewable Energy Resources (Parliament Resolution No. 160-XVI, 12 July 2007[26]); ANRE's Decision No. 321, 22 January 2009, Official Monitor No. 45–46 of 27 Febuary 2009[27]).
- *Guarantee Of Origin (GOO):* It specifies the RES from which electricity was produced, location of production, date of issue, type of renewable, power generation installed capacity, quantity of electricity produced and allows the manufacturer to demonstrate the origin of RES-E. It is issued for RES of capacity higher than 10 kW. In the case of waste incineration installation network operator grants GOO based on the amount of electricity produced. (Regulation on the GsOO of Electricity Generated from Renewable Electric Energy and Biofuel (ANRE's Decision No. 330 from 03 April 2009, Official Monitor No. 99-100 of 5 June 2009); same Law as above).

---

22. http://www.konkurentsiamet.ee/?id = 14705.
23. http://www.envir.ee/67244.
24. http://www.mkm.ee/.
25. http://elering.ee/.
26. http://lex.justice.md/index.php?action = view&view = doc&lang = 1&id = 324901.
27. Methodology for the determination, approval and application of tariffs for the RES-E and for bio-fuel. Official Monitor No. 45-46/27.02.09.

- *Free zones:* They concern investments for RES aiming to accelerate socio-economic development of certain territories and the country as a whole (Law 451-XV/ 30 June 2001, published on 04 March 2011, in Official Gazette No. 34–36, Article No. 7, issued on 18 February 2011[28]).

The IN includes: Ministry of Economy and Trade;[29] Ministry of Environment;[30] National Agency for Regulation of Energy (ANRE).[31]

[I]     Romania

The policy mixture for RES includes:

- *GOO:* As in Bulgaria (GD 219/2007 – Official Journal Part I, No. 200/23 March 2007- and GD 1215/2009 – Official Journal No. 748/3 November 2009). This GD was followed by the Procedure regarding the issuance of the GOO (GD 1461/2008) and GD 1215/2009).
- *Support scheme during the period 2010–2023:* It promotes high-efficiency cogeneration based on useful heat demand and is applied for: (a) Producers of electricity and thermal energy in high efficiency cogeneration; (b) electricity consumers; (c) suppliers of electricity; (d) network operators; (e) the administrator of supporting the scheme. The maximum installed capacity of cogeneration plants able to receive the support scheme is 4000 MW. The bonuses unit (for the first year of the award, 2010) were: (i) RON 157,53/MWh (EUR 37,41/MWh) for plants using mainly natural gas in the transport network; (ii) RON 166,24/MWh (EUR 39,48/MWh) for plants using mainly natural gas in the distribution network; (iii) RON 179,12/MWh (EUR 42,54/MWh) for plants using mainly solid fuel (GD 219/2007 – Official Journal Part I no. 200/23 March 2007 and GD 1215/2009 – Official Journal No. 748/ 3 November 2009).
- *Quota combined with Green Certificate system:* There is a mandatory annual quota increase for RES. At the end of each year, distribution companies have to deliver a certain amount of 'Green Certificates (GCs)' corresponding with the annual quota.[32] RES-E producers will receive for every 1 MWh produced and delivered in the electric power network from: (i) new hydroelectric stations/groups with a maximum output of 10 MW, 3 GCs; (ii) refurbished hydroelectric stations/groups with a maximum output of 10 MW, 2 GCs; (iii) biomass, biogas, bio liquid, waste fermentation gas, geothermal power and associated combustible gases – 3 GCs; (iv) solar power – 6 GCs; (v) wind energy 2 GCs, until 2017, and 1 GC, starting from 2018. RES-E producers will receive 1 GC for every 2 MWh delivered in the electric power network from

---

28. http://lex.justice.md/index.php?action = view&view = doc&lang = 1&id = 324901 and http://lex.justice.md/md/337739/.
29. http://www.mec.gov.md/.
30. http://www.mediu.gov.md/index.php/en/.
31. http://www.anre.md/index.php?vers = 3.
32. http://investeast.ro/renewable_energy_in_romania.pdf.

hydroelectric stations with an installed power between 1 and 10 MW (Law 220/2008 – Official Journal No. 743/3 November 2008; Republished: Official Journal No. 577/13 August 2010); modified by Law 139/2010 – Official Journal No. 474/9 July 2010; and GEO 88/2011 – Official Journal No. 736/19 October 2011).

The IN includes: Ministry of Economy, Trade and Business Environment;[33] Ministry of Environment and Forests;[34] National Environmental Protection Agency;[35] Natural Agency of Mineral Resources;[36] Romanian Energy Regulatory Authority;[37] Romanian Transmission and System Operator – Transelectrica.[38]

[J]   Russia

The RES policy mixture includes:

- *Premium scheme:* It is determined by the Federal Service for Tariffs and added to the wholesale electricity market price, forming the final price of RES-E (Boute, 2012). Electricity buyers finance the premium scheme in proportion to the amount of electricity that they are obliged to purchase on the wholesale market (IFC, 2011b) (Federal Law No. 35-FZ, issued on 26 March 2003, 'On the Electric Power Industry'; amended by Federal Law No. 250-FZ issued on 4 November 2007; Decree of the Government of the RF No. 426, issued on 3 June 2008 (modified by Decree No. 58, issued on 5 February 2010); Regulation 'on the Certification of a Production Installation Using RES and on the Administration of the Register of Certified Production Installations', approved by Protocol of the Supervisory Board of the Market Council of 3 October 2008).
- Compensation of the grid connection costs of RES-e facilities with capacity under 25 MW (IFC, 2011b; Boute, 2012) (Same laws, decrees as above).
- *Priority:* Grid companies prioritize RES-e when covering losses on the transmission grid (IFC, 2011b; Boute, 2012) (Same laws, decrees as above).
- *Certificate:* Installations of RES-E must be certified as 'renewable energy installations' so as to receive support under the 'premium' and 'certificates' scheme (Boute, 2011). Large hydropower installations fall under the definition of renewable energy (Federal Law 'On the Electric Power Industry') but are not entitled to receive support (Boute, 2011). The certification of renewable energy installations requires also their inclusion in the '*Scheme for the Location of Electricity Production Installations Using RES*' elaborated by the Ministry of

---

33. http://www.minind.ro/.
34. www.mmediu.ro/.
35. www.anpm.ro.
36. www.anrm.ro.
37. www.anre.ro.
38. www.transelectrica.ro.

Energy (Boute, 2011). The Ministry of Energy specifies the location and type of renewable energy generation installation that it decides to support (IFC, 2011b; Boute, 2011; 2012). Certificates are valid for a period of three years starting from their date of issue (Boute, 2011). Their life cycle includes their issuance, transfer and cancellation (Boute, 2011). The Russian authorities did not plan to create a secondary market for tradable certificates (Boute, 2011) (Order No. 187/17 November 2008; Order of the Russian Government No. 1166-r/ 18 August 2009).

- *Capacity based scheme:* RES are promoted through the remuneration of the installed capacity of renewable energy generating facilities (IFC, 2011b). On 6 December 2011, Russia amended the Federal Electricity Law providing the possibility to support RES by concluding with investors on 'Agreements for the Delivery of Capacity' (IFC/GEF, 2012; Boute, 2012). These Agreements are long-term regulated capacity contracts of certain installations selected by the Government in the context of the corporate restructuring of the former quasi-monopolist RAO UES providing also remuneration at regulated prices. Actually, these agreements renumerate the readiness of electricity generating facilities to produce electricity. Investors commit to construct a certain type of production installation of a certain capacity, at a certain location in return for long-term regulated tariffs (IFC/GEF, 2012). They also commit to maintain their installations ready for producing electricity (Boute, 2012). If electricity producers fail to guarantee the availability (dispatchability) of their installations, then remuneration according to capacity to which they are entitled is decreased by specific coefficients (Boute, 2012). Regulated RES tariffs are determined annually according to Principles of Price Regulation in the Electricity Sector (Decree of the Government of the RF No. 1178 of 29 December 2011) (IFC/GEF, 2012). Regional tariff authorities can adopt tariffs with duration longer than five years only after the approval of the Federal Service for Tariffs, the Ministry of Energy and the Ministry for Economic Development according to the same Principles (IFC/GEF, 2012) (Federal Law No. 401-FZ, issued on 28 December 2010, amended on 6 December 2011).

The IN includes: Ministry of Energy;[39] Ministry of Natural Resources and Environment;[40] Ministry of Regional Development;[41] Sberbank;[42] Federal Tariff Service or Federal Agency for Tariffs;[43] Commission on modernization and technological development of the Russian economy.

---

39. http://minenergo.gov.ru/aboutminen/leaders/.
40. http://www.mnr.gov.ru/english/.
41. http://www.minregion.ru/.
42. http://sberbank.ru/en/corporatecustomers/carbonfinance/.
43. http://www.fstrf.ru/eng.

## [K] Serbia

The policy mixture for RES includes:

- *Privileged producers:* It refers to producers of electrical/thermal energy who use New RES or waste for energy production. They are entitled to: (i) priority status in the energy market against other producers provided that their offers are under equal conditions and (ii) usage of subventions, tax, customs, and other exemptions (Tesic M. et al., 2011) (Energy Law - OJ RS 84/2004; Decree - OJ 72/2009).
- *FITs:* (Decree - OJ 99/2009, 1 December 2009).
- *PPAs:* for 12 years (CMS, 2013).
- *Green certificates scheme* (Updated Energy Law - OJ 57/2011 - 01 August 2011).

The IN for RES includes: Ministry of Energy, Development and Environmental Protection of Republic of Serbia;[44] Ministry of Natural Resources, Mining and Spatial Planning;[45] Energy Agency of the Republic of Serbia[46] (AERS); Electric Power Industry of Serbia.[47]

## [L] Turkey

The RES policy mixture consists of:

- *FITs:* They were launched in 2007, ranging from EUR 5 cents/kWh and up to EUR 5.5 cents/kWh. In 2011, the establishment of *'Renewable Energy Support Mechanism'* was applied to plants commissioned between 2005 and 2015, and enabled them to benefit from FITs for 10 years (Law No. 5346 - Official Gazette: dated 18 May 2005 and numbered 225819; amended by Law No. 6094 - Official Gazette: dated 8 January 2011 - numbered 27809).
- *Certification of RES* (similar to GOO in EU Directive) (same Law as above).
- *Grid-accession priorities* (same Law as above).
- *Exemption from licensing obligations and setting up a company*: It concerns individual and corporate entities that are to build electricity generation facilities from RES having maximum installed capacity of 500 kW. Private sector entrepreneurs were allowed to build and operate power plants by receiving a license from Energy Market Regulatory Authority (EMRA) (Law No. 4628 - Official Gazette: dated 3 March 2001, numbered 24335) amended by Law No. 5784 - Official Gazette: dated 26 July 2008, numbered 26948).
- *Allowed construction of RES plants:* They are allowed with permissions obtained from the Ministry of Forestry and Water Affairs and the Ministry of

---

44. http://www.merz.gov.rs/en.
45. http://www.mprrpp.gov.rs/en/.
46. http://www.aers.rs/Index.asp?l = 2&a = 540&ted = &ed = &tp = &id = &idag = &tvid = .
47. http://www.eps.rs/Eng/index.aspx.

Environment and Urbanism to be constructed in protected regions (such as national and natural parks, natural monuments, protected regions, conserved forestry, wildlife development zones, special environmental protection zones and natural protected areas) (Law No. 6094).

### [M]  Ukraine

The set of policy instruments for the RES exploitation includes:

- *'Green' tariff:* It is a special tariff for the purchase of electricity produced at power plants using alternative energy sources (except for blast-furnace and coke gas, and using hydro energy – produced by SHPPs). Electricity produced in such way can be sold directly to consumers, and the wholesale market is obliged to pay the 'green' tariff for this type of produced electricity and not sell it at contractual prices directly to consumers or energy supplying companies[48] (Law No. 601-VI of 25 September 2008; amended by Law No. 1220-VI of 01 April 2009[49]).
- *Minimum share of raw materials, fixed assets, works and services necessary for the construction of the power plant that must be of Ukrainian origin.* Starting from year 2012, this share of national material and labor must be more than 30% of the total construction costs, and starting from year 2014, it must exceed 50%. For solar power plants the national component in the production of solar modules needs to exceed the 30% starting in year 2012. The exact procedure for the determination and control of these shares has not been approved to date[50] (Same Law as above).
- *Obligatory volume of production and use of bio and mixed motor fuel* (Law No. 1391-VI of 21 May 2009).
- *GOO:* (Law No. 5485-VI of 20 November 2012[51]).
- *GIS:* Facilitation for the use of unexploited energy sources like waste energy (Resolution of the CM of Ukraine of 16 July 2008 No. 642;[52] Resolution of the CM of Ukraine of 30 October 2008 No. 1369-p;[53] Resolution of the CM of Ukraine of 16 September 2009 No. 1034[54]).

The RES IN is formed by: Ministry of Energy and Coal Industry;[55] Ministry of Ecology and Natural Resources;[56] National Electricity Regulation Commission.[57]

---

48. http://zakon2.rada.gov.ua/laws/show/601-17.
49. http://zakon2.rada.gov.ua/laws/show/1220-17.
50. http://ukrainian-energy.com/en/energy_legislation/articles/details/177.
51. http://zakon2.rada.gov.ua/laws/show/5485-17.
52. http://zakon2.rada.gov.ua/laws/show/642-2008-%D0%BF.
53. http://zakon.nau.ua/doc/?uid=1095.3558.0.
54. http://zakon1.rada.gov.ua/laws/show/1034-2009-%D0%BF.
55. http://mpe.kmu.gov.ua.
56. http://www.menr.gov.ua/.
57. http://www.nerc.gov.ua/.

## §8.03  LEAP MODEL

The Long range Energy Alternatives Planning System (LEAP), developed by SEI's U.S. Center, is an integrated modeling software tool, widely used for energy policy analysis and climate change mitigation assessment (SEI, 2012b). It allows to: (i) track energy consumption, production, and resource extraction in all sectors of an economy; (ii) account for both energy sector and non-energy sector Greenhouse Gas (GHG) emission sources and sinks (iii) analyze emissions of local and regional air pollutants, making it well-suited to studies of the climate co-benefits of local air pollution reduction.

LEAP can serve as a historical database demonstrating the evolution of an energy system and a forward-looking, scenario-based tool. It relies on simple accounting principles, and many of its aspects are optional, keeping its initial data requirements relatively low (SEI, 2012b). It is possible to use it for a wide range of scales, from cities and states to national, regional, and even global applications. It has been adopted by thousands of organizations in more than 190 countries worldwide, including government agencies, academics, non-governmental organizations, consulting companies, and energy utilities (SEI, 2012b).

### [A]  Use of LEAP

For each one of the 11 countries a LEAP dataset was prepared (tree representing the energy system of the country along with historical data (current account)) during the PROMITHEAS-4 project. All historical data were sought from national and international official sources.

The next step was to determine the key assumptions for the evolution of the most important drivers, but simultaneously to adopt a common approach for all countries. The time evolution for: (i) population was based on the projections of the Department of Economic and Social Affairs of the United Nations (UN, 2011); (ii) National real GDP on the projections of the International Monetary Fund (IMF) (IMF, 2012). Depending on the availability of historical data the growth of the variable 'Final energy intensity' or 'Total energy' of an economic national sector was linked to the growth of the 'GDP real'. The use of GDP real over GDP nominal was preferred for removing the effect of inflation and being able to compare the outcomes among all countries. The fuel share for each sector remained as it was since this analysis focuses on the currently implemented policy mixtures that had already imposed changes.

*Figure 8.2  LEAP Interface (Figure from Running the Model)*

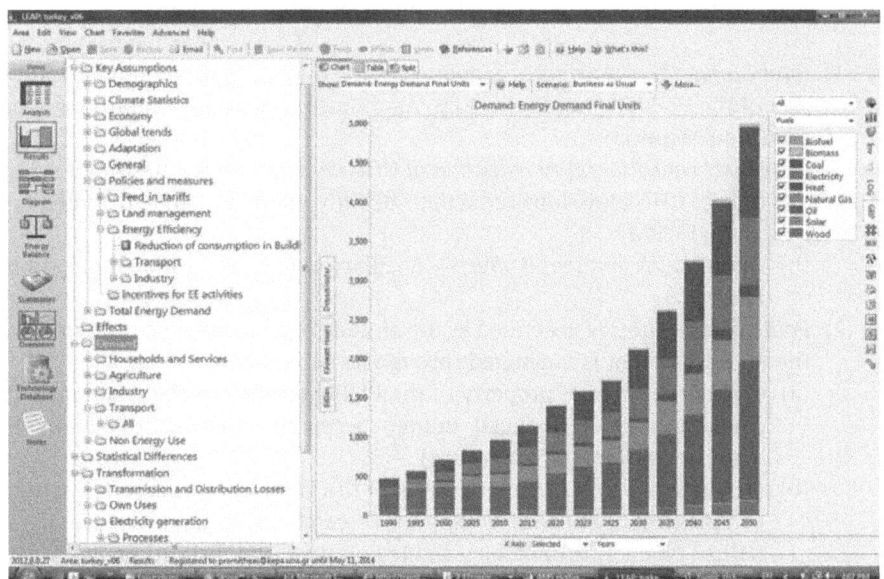

## §8.04  AMS METHOD

Each policy mixture will be assessed for its performance under the criteria/sub-criteria of the AMS method which is the combination of three standard multi-criteria methods: the Analytical Hierarchy Process (AHP), the Multi-Attribute Utility Theory (MAUT), and the Simple Multi-Attribute Ranking Technique (SMART) (Konidari and Mavrakis, 2007; 2006). AMS is developed for evaluating climate policy instruments (PI) or relevant Policy Mixtures (PM) and with suitable modification for evaluating their interactions as well.

The AHP procedure provides values for the weight coefficients of criteria /sub-criteria. The other two – MAUT/SMART – provide normalized grades for the performance of PIs or PMs under the selected criteria/subcriteria. The MAUT procedure is applied when the user of the method has available and credible data under the same criterion/sub-criterion for all the evaluated PIs/PMs, while the SMART procedure when – due to the absence of data – grades are based on user's judgment and/or experts' opinion.

AMS consists of four basic steps: (1) creation of criteria-tree; (2) determination of weight coefficients for criteria/sub-criteria; (3) grading for the performance of the PI/PM under a criterion/sub-criterion; (4) collection of the previously produced grades and formation of the aggregate grade for each evaluated PI/PM. Consistency and robustness tests are performed within the relevant steps.

For the first step, the criteria-tree is that of previous applications (Figure 8.1) (Konidari and Mavrakis, 2007; 2006). The goal of climate PIs/PMs (first level) is to be effective in mitigating climate change through GHG emissions reductions and confronting expected adaptation needs. The second level includes three criteria: environmental performance, political acceptability, and feasibility of implementation:

(1) *Environmental performance* is defined as the overall environmental contribution of the PI/PM towards the goal. Assessment under this criterion is based on the sub-criteria:
   (a) *Direct contribution to reduction of GHG emissions* – synthesis and magnitude of GHG emissions reductions directly referred to and attributed only to the PI/PM.
   (b) *Indirect environmental effects* – ancillary outcomes attributed only to the PI/PM.
(2) *Political acceptability* is defined as the attitude of all involved entities towards the PI. Assessment is facilitated through its sub-criteria:
   (a) *Cost effectiveness* – property of the PI/PM to achieve the goal under the perspective of a financial burden acceptable and affordable by the involved entities (target groups).
   (b) *Dynamic cost efficiency* – property of the PI/PM to create, offer or allow compliance options that support research projects, incremental and radical pioneer technologies and techniques, and institutional or organizational innovations leading to GHG emission reductions and lessening the impacts of climate change impacts.
   (c) *Competitiveness* – capacity of the entity to compete, under the particular PI/PM, via price, products or services with other entities and maintain or even increase the magnitude of specific indicators describing its financial performance.
   (d) *Equity* – fairness of the PI/PM in distributing emission rights, compliance costs and benefits among entities (countries/sectors) for accomplishing GHG emission reductions and handling climate change impacts. For international instruments two types of equity are considered, *international (global)* and *inter-generational*. The first concerns equity among countries, while the second refers to the time horizon of the PI/PM for confronting climate change. National PIs/PMs are examined for *intra-country* equity, which is equity within the country. Intra-country equity is divided into sector and social equity. *Sector equity* is the perceived fairness between different national sectors regarding the climate change burden. *Social equity* is the perceived equity between different groups of society.
   (e) *Flexibility* – the property of the PI/PM to offer a range of compliance options and measures that entities are allowed to use in achieving reductions under a time frame adjusted according to their priorities.
   (f) *Stringency for non-compliance and non-participation* – level of rigidity determined by provisions of the PI/PM towards emitters that failed to comply or did not participate to its implementation.

(3) *Feasibility of implementation (or enforcement)* is defined as the aggregate applicability of the PI/PM linked with national infrastructural (institutions and human resources) and legal framework. Assessment is based on:
   (a) *Implementation network capacity* – ability of all national competent parties to design, support, and ensure the implementation of the PI/PM. The capacity of the network is based on its *trained personnel, technological infrastructure, credibility* and *transparency*. The *trained personnel* concern the national human resources capable in supporting implementation of the PI/PM. *Technological infrastructure* is the set of available technologies and techniques within the country that can be used for supporting implementation. *Credibility* is defined as the accuracy and consistency that characterize its activities, mainly measurements and elaboration of data necessary for implementation, promotion, and steering of national compliance efforts. *Transparency* is defined as the openness of the implementation network towards target groups in providing them with clear information for the implementation of the PI/PM and methods of operation.
   (b) *Administrative feasibility* – aggregate work exerted by the regulatory implementation network during the enforcement of the PI/PM.
   (c) *Financial feasibility* – property of the PI to be implemented with low overall costs by the pertinent regulatory authorities.

For the second step, the values for the weight coefficients of the aforementioned criteria/sub-criteria are those calculated in previous work (Konidari and Mavrakis, 2007). The consistency test was also performed for these values with very good results (Konidari and Mavrakis, 2007).

The third step will be presented analytically in the following session. For the fourth step a grade (commonly measured performance) – determined in third step – of the assessed PI/PM for a certain sub-criterion is multiplied with the respective weight coefficient of the sub-criterion. All products (concerning all sub-criteria) are added and form the grade of the criterion that is supported by these sub-criteria. The sum of these products is the grade of the PI/PM under the criterion. This criterion grade is multiplied with the respective weight coefficient of the criterion. All new products are added and form the final grade, which expresses the effectiveness of the evaluated PI/PM. Calculations will be done with the software Clim AMS, which is developed to facilitate work with AMS.

## §8.05 EVALUATION

All 11 policy mixtures are evaluated against the described criteria and their respective sub-criteria.
   *Criterion 1: Environmental Performance*
   *Direct Contribution to GHG emissions:* The outcomes of the LEAP model are used for evaluating the performance of the policy mixtures under this sub-criterion. The

policy mixture that has achieved the highest penetration of RES so as to reach the set national RES target is the one that contributes more to the reduction of GHG emissions. The calculations and the results are presented in Table 8.1. Calculations based on LEAP outcomes were needed since the officially presented achievements of previous years did not coincide with those of the available historical data.

*Indirect environmental effects:* the total amount of the total environmental effects is provided by LEAP outcomes (Table 8.3).

*Cost efficiency:* Based on the work performed in PROMITHEAS-4, an index was developed so as to understand the benefit-cost gains that the implementation of a policy instrument has on the target groups. The classification of the marginal abatement costs for certain technologies promoted under certain policy instruments was used to specify this index.

For each of these 11 countries a mean Cost efficiency index is calculated depending on the implemented policy instruments (FITs/emission trading: −0.25 pseudo-monetary units per $tCO_2eq$, Quota obligation/obligator purchase: −0.75 pseudo-monetary units per $tCO_2eq$.

*Figure 8.3   The AHP Hierarchy (Konidari and Mavrakis, 2006)*

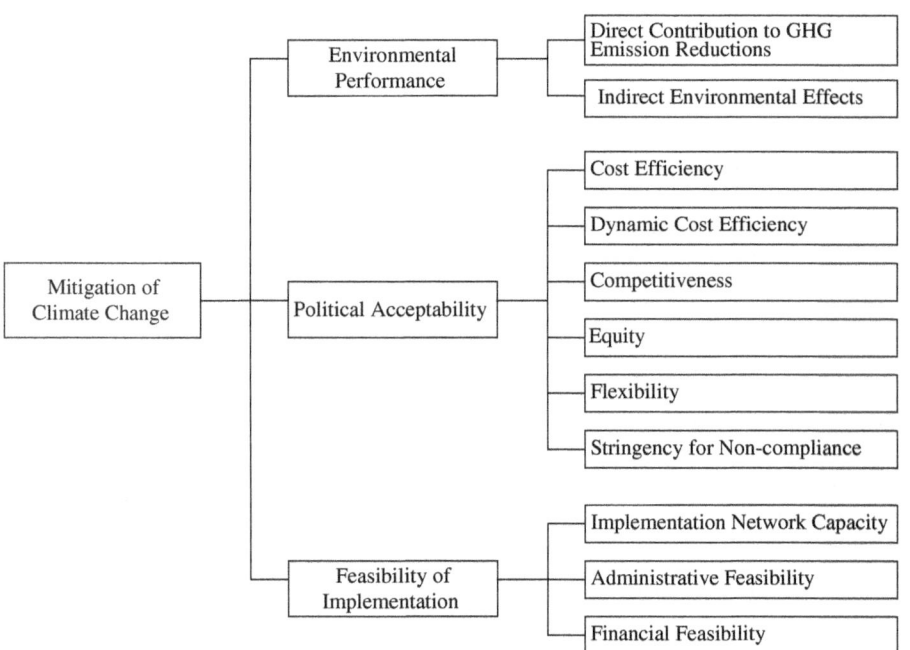

Chapter 8: Evaluation of Implemented Policy Mixtures for RES      §8.05

Table 8.1  Presentation of Recorded, Targeted and Estimated Share of RES for the Eleven Emerging Economies (Eurostat, 2012; Energy Community, 2012; Willems, S. and Anakhasyan, E., 2011; UNECE, 2011; IFC, 2011b; IPA Energy + Water Economics, 2010; European Commission, 2009). N/A – Not Available, E – Estimation (**Share of RES in Electricity Generation)

|  | Year 2009 – (%) Share of RES in | | | | Year 2020 – (%) | | Difference to Cover between Recorded Share of Base Year 2009 and Target Share of Year 2020 (A) | Estimation of LEAP Based on Historical Data for RES Share in Year 2009 | Estimation of LEAP for RES Share in Year 2020 | Difference Covered between Share of Base Year 2009 and Share of Year 2020 According to LEAP Calculations (B) | (B)-(A) | MAUT scale |
|---|---|---|---|---|---|---|---|---|---|---|---|---|
|  | Gross Final Energy Consumption | Gross Electricity Consumption (*Electricity Generation) | Heating and Cooling | Transport | Target - Share of RES in Gross Final Energy Consumption | Target - Share of RES in Electricity Generation |  |  | (%) |  |  |  |
| Albania | 31.2 | N/A | N/A | N/A | 38 |  | 6.8 | 26.69 | 44.9 | 15.22 | 8.42 | 95.24 |
| Armenia | 3.3 (2008) | N/A | N/A | N/A | 20 (E) |  | 16.7 | 8.98 | 13.24 | 4.25 | -12.40 | 0.00 |
| Azerbaijan | 0.4 (2007) | N/A | N/A | N/A | 9.7 |  | 9.3 | 3.17 | 1.93 | -1.24 | -10.50 | 8.69 |
| Bulgaria | 11.9 | 11.3 | 21.0 | 0.6 | 16 |  | 4.1 | 19.12 | 21.3 | 2.18 | -1.90 | 48.03 |
| Estonia | 23.0 | 6.1 | 41.8 | 0.2 | 25 |  | 2 | 24.5 | 36.01 | 11.46 | 9.46 | 100.00 |
| Moldova | 11.9 | N/A | N/A | N/A | 17 |  | 5.1 | 16.23 | 14.65 | -1.58 | -6.68 | 26.17 |
| Romania | 22.4 | 30.9 | 26.4 | 1.6 | 24 |  | 1.6 | 24.75 | 26.45 | 1.70 | 0.10 | 57.18 |

181

| | Year 2009 – (%) Share of RES in | | | | Year 2020 – (%) | | Difference to Cover between Recorded Share of Base Year 2009 and Target Share of Year 2020 (A) | Estimation of LEAP Based on Historical Data for RES Share in Year 2009 | Estimation of LEAP for RES Share in Year 2020 | Difference Covered between Share of Base Year 2009 and Share of Year 2020 According to LEAP Calculations (B) | (B)-(A) | MAUT scale |
|---|---|---|---|---|---|---|---|---|---|---|---|---|
| | Gross Final Energy Consumption | Gross Electricity Consumption (*Electricity Generation) | Heating and Cooling | Transport | Target–Share of RES in Gross Final Energy Consumption | Target–Share of RES in Electricity Generation | | | (%) | | | |
| Russia | N/A | 1.5* (2008) | N/A | N/A | - | 4.5 | 3 | 1.97 | 1.92 | -0.05 | -3.05 | 42.77 |
| Serbia | 21.2 | N/A | N/A | N/A | 27 | | 5.8 | 15.41 | 9.20 | -6.21 | -12.01 | 1.78 |
| Turkey | 15.6 (2005) | N/A | N/A | N/A | 25** (E) | 30 (2023) | 9.4 | 13.51 | 12.18 | -1.32 | -10.72 | 7.69 |
| Ukraine | 5.5 | N/A | N/A | N/A | 11 | | 5.5 | 5.37 | 3.10 | -2.27 | -7.78 | 21.13 |

*Table 8.2  Matrix of Policy Instruments for RES per Country*

| | Countries | | | | | | | | | | |
|---|---|---|---|---|---|---|---|---|---|---|---|
| | Albania | Armenia | Azerbaijan | Bulgaria | Estonia | Moldova | Romania | Russia | Serbia | Turkey | Ukraine |
| | *Policy instruments* | | | | | | | | | | |
| FITs | X | X | X | X | | X | | | X | X | X |
| Premium scheme | | | | | X | | | X | | | |
| Capacity based scheme | | | | | | | | X | | | |
| Obligatory purchase | | | | X | X | | | | | | |
| Tax exemptions/ reductions | X | | X | | | | | | | | X |
| Tradable Green certificates | | | | | | | X | | X | | |
| Quota obligation | | | | X | | | X | | | | |
| Obligatory production | X | | | | X | | | | | | X |
| Emission trading schemes (JI, CDM,GIS) | | | | X | X | | X | | | | X |
| *Rules – influencing mechanisms* | | | | | | | | | | | |
| PPAs | X | X | | X | | | | | X | | |

| | Countries | | | | | | | | | | |
|---|---|---|---|---|---|---|---|---|---|---|---|
| | Albania | Armenia | Azerbaijan | Bulgaria | Estonia | Moldova | Romania | Russia | Serbia | Turkey | Ukraine |
| Penalties | | | | X | | | | | | | |
| Investment aid (Grants, soft loans, Subsidies) | | | | | X | | X | X | | | |
| Privileged producers/access to grid | X | | | X | | | | | X | X | |
| Minimum share of national sources | | | | | | | | | | | X |
| Certificate/Guarantee of origin | | | | X | X | X | X | | | X | X |
| Certificate | | | | | | | | X | | | |
| Free zones, exemption of licensing obligations, permissions to | | | | | | X | | | | X | |
| ERUs, CERs linked with RES | | X | X | | | | | | | | |

Table 8.3 Indicators and Grades for the Evaluation

| | Difference between Base Year and 2020 in Emissions for CO, NOx, Non Methane Volatile Organic Compounds, SO$_2$ (LEAP Outcomes in MtCO$_2$eq) | MAUT Scale | Cost Efficient Index (Pseudo-Monetary Unit/tCO$_2$eq) | MAUT Scale | Annual Difference of Energy from RES Consumed in Base Year and 2020 per capita (Million Gigajoule per Capita per Year) | MAUT Scale |
|---|---|---|---|---|---|---|
| Albania | 0.086 | 95.47 | 0.416 | 66.400 | +0.809 | 34.72 |
| Armenia | 0.101 | 94.68 | 0.250 | 0.000 | +0.070 | 4.32 |
| Azerbaijan | 0.322 | 83.05 | 0.250 | 0.000 | -0.035 | 0.00 |
| Bulgaria | 0.003 | 99.84 | 0.500 | 100.00 | +0.544 | 23.82 |
| Estonia | 0.199 | 89.52 | 0.500 | 100.00 | +2.396 | 100.00 |
| Moldova | 0.000 | 100.00 | 0.250 | 0.000 | +0.102 | 5.63 |
| Romania | 0.880 | 53.68 | 0.416 | 66.400 | +0.521 | 22.87 |
| Russia | 1.100 | 42.11 | 0.250 | 0.000 | +0.278 | 12.87 |
| Serbia | 0.556 | 70.74 | 0.250 | 0.000 | -0.009 | 1.07 |
| Turkey | 1.600 | 15.79 | 0.250 | 0.000 | +0.110 | 5.97 |
| Ukraine | 1.900 | 0.000 | 0.375 | 50.00 | -0.019 | 0.66 |

*Dynamic cost efficiency:* there are no available data from LEAP so the SMART procedure is used. Emphasis is given to the range of RES types that are promoted and if research and innovation for RES are encouraged.

*Albania:* Only biomass and hydro are used in the national energy mix. Solar energy is not used because of unaffordable prices for the private sector and households (AgriPolicy, 2009b). Solar systems are almost entirely imported; mostly bought by few specialized companies (AgriPolicy, 2009b). That is why they are considered expensive, and used only as partial solution (hot water for technical use, not heating) (AgriPolicy, 2009b). No research efforts for RES technologies are promoted.

*Armenia:* Mainly hydro is used while biogas and wind have vey low shares. Investment in research and innovation remains at relatively modest levels (European Commission/High representative of the European Union for Foreign Affairs and Security Policy, 2012). In 2010, Armenia reached its target only for SHPPs according to national market monitoring reports published by the PSRC, and it is not expected to reach its targets for 2025 (IFC, 2012). PSRC estimations show that total cumulative investment in SHPPs will reach the amount of USD 190 million in 2015, while investments in biogas, wind, and other RES technologies will be negligible (IFC, 2012). The usage of solar energy in households and in industry (e.g., agricultural product processing enterprises, middle and average enterprises) is low due to limited access to funds, promotional financial mechanism and institutional support (UNDP/GEF, 2012).

*Azerbaijan:* Very low share of hydro, wind, and biomass in the energy mix. Main reasons are: (i) no elaborated incentives for attracting investments; (ii) lack of technical capacity to support any investment interest and (iii) lack of a comprehensive, clear tariff methodology. No support of such technologies under the CDM due to low incentives for companies. No research support.

*Bulgaria:* Biomass, solar, hydro, wind, and bio-fuels are used. Innovation is at low levels. According to the Bulgarian 'Energy Strategy until 2020'[58] financial support for research will be sought 'through facilitation of investors' access to scientific development works, specialized credit lines and facilities from European funds and programs'. No further action for this.

*Estonia:* Hydro, wind, biomass, municipal waste, and biogas are part of the national energy mix. The renewable subsidy program allows the development of research projects related to the investigation of RES potential and other innovative technologies.

*Moldova:* Wood, vegetal waste, and hydro are used with the last two at very low shares compared to wood. No promotion or support of innovation in RES technologies.

*Romania:* Biomass, hydro, wind, and bio-fuels are used. Research on new RES technologies is limited (Patlitzianas K., Karagounis K., 2011; Mihailescu L., 2009). The use of solar energy and biomass in Romania is still mostly in the inception phase with a few experimental projects, which were set up across the country (Nitoiu M, 2011).

*Russia:* Hydro and biomass are mainly used, while biogas and solar have much smaller shares. The country has low rates of technological adoption (137th in the

---

58. http://www.mi.government.bg/files/useruploads/files/epsp/23_energy_strategy2020%D0%95ng_.pdf.

world) (World Economic Forum, 2013). It is indicative that Research and Development accounted for only 3% of the total investment in terms of FDI projects (Ernst & Young, 2012b). Also, the vast availability of fossil resources in the country did not offer incentives to increase deployment of RES (UNECE, 2010).

*Serbia:* Utilization of RES is limited to hydropower plants, non-commercial use of biomass and very low geothermal energy, while hydropower is the only RES used for electricity generation. Production of bio-ethanol and bio-diesel is still small. Almost all technologies for biomass energy conversion have been applied, but some of them were installed more than 20 years ago and are by now out of operation (Bogunovic A. and Bogdanov N., 2009). Due to low electricity prices a significant number of owners considered as much cheaper to use electricity for heating than biomass and manufacturers did not invest in the development of modern solid fuel firing boilers (Bogunovic A. and Bogdanov N., 2009). The level of technological improvements in the waste management and recycling sector is low and collecting and sorting household waste is not the practice in this area (Embassy of Denmark in Belgrade, 2010). Promotion of relevant research issues is poor. The realization of Research and Development projects is impossible without modern and sophisticated equipment (Republic of Serbia, 2010). The laboratories are not well equipped so it is necessary to outline a program of modernization for laboratories (Republic of Serbia, 2010).

*Turkey:* Hydro, biomass, wood, geothermal, and wind are used. Electricity generation from hydropower is already a mature technology. No research efforts for RES technologies are promoted.

*Ukraine:* Biomass, hydro, and wind are used. In 2011, the installed capacity of RES power plants increased more than twice due to the new FITs for RES, but the Law on the 'green tariffs' did not stimulate the construction of new RES plants (Deloitte, 2012; Trypolska G., 2012; Energy Community Regulatory Board, 2011). The RES share in total energy generation for year 2011 was 0.2%, which is relatively small, indicating huge potential and little to no competition (Deloitte, 2012). FITs were applied only for wind, biomass, solar, and small hydropower plants (EBRD/GEF, 2011). Hydropower plants below the threshold of 10 MW are considered as SHPPs and incentivized through the FITs. The green tariff does not support biogas, all types of biomass or geothermal energy (EBRD/GEF, 2011). The majority of hydropower plants remains outdated and requires modernization (Deloitte, 2012). In 2010, only 49 out of 150 small and micro plants were actually operating (OECD, 2012). When the green tariffs were first introduced in 2009, large plants providing electricity at market rates switched to the subsidized rate, implying the adjustment of incentives so as to support the creation of new capacity (OECD, 2012).

Grades: Albania – 3, Armenia – 3, Azerbaijan – 2, Bulgaria – 5, Estonia – 6, Moldova – 3, Romania – 4, Russia – 2, Serbia – 3, Turkey -2, Ukraine – 3.

*Competitiveness:* There are no available data from LEAP, so the SMART procedure is used. Emphasis is given to the attractiveness of the country for RES investors and the difficulties that they face.

*Albania:* Investments in RES projects from foreign private capitals are not mentioned in relevant national reports.[59] Particularly for promoting wind power, there is lack of consecutive measurements of the velocity and duration of the wind (AKBN, 2010). Various foreign companies willing to invest in this type of RES need to make a prior assessment of the records of the wind velocity and duration (AKBN, 2010).

*Armenia:* Foreign commercial banks have little incentive to finance in Armenian RES projects (either as an individual investment or as a CDM one) due to existing investment risks, modest returns, and high demand for funds compared to safer proven markets (USAID, 2010). The cost to develop a wind project in Armenia – during 2007–2008 – was at EUR 1800–1900/kW (compared to EUR 1300–1500/kW in Europe) (USAID, 2010). International financial institutions (i.e., EBRD, IFC, ADB, etc.) frequently offer loans at commercial terms plus risk coverage premiums, but in this manner RES projects and particularly wind projects are economically not profitable at current terms (USAID, 2010). Usually investors do not prefer to set RES projects under the CDM since Armenian CERs seem nearly irrelevant because of the low-carbon factor in generation mix (USAID, 2010). Also, commercial banks are unwilling to consider the potential income from carbon sale as a serious factor in long-term prediction and economic forecasting of wind projects (USAID, 2010). Current FIT structure does not secure reasonable profitability even for SHPPs (IFC, 2012).

Most RES investors have been funded through International Donor Organizations (IDO) at social and state owned sites and not from commercial incentives, including state grants (Danish Energy Management/SolarEn LLC, 2009). Usually soft commercial loans are negotiated by the government with these IDO (World Bank, European Bank for Reconstruction and Development, etc.) (Danish Energy Management/SolarEn LLC, 2009). Investors face also legislative issues also. A new application is submitted 30 days prior to the expiration of the license (tracking the life of the power purchase agreement) (USAID, 2011b). However until the application is submitted, reviewed by the regulator and approved, there is no mechanism to guarantee investors that the agreement will be renewed, at which tariff levels it will be renewed, and the length of the renewal (USAID, 2011b). Another problem that potential investors face is that turbines imported for SHPPs are free of VAT and customs duties, but the law does not specifically mention that generators and other related control components are also exempted (Danish Energy Management (A/S), 2011).

Armenia can be competitive in PV technologies development due to the existence of a wide variety of siliceous raw material from various sources and morphology in its territory (Danish Energy Management/SolarEn LLC, 2009).

*Azerbaijan:* Its economy depends strongly on its oil and gas sector, which is why it has successfully attracted large investments for this sector due to a favorable operating environment for investors (EBRD, 2010). However, the government is interested in developing the national EE and RES potential also and particularly

---

59. Few RES pilot projects are developed by donors (such as World Bank, EBRD, KfW, UNDP). Private sector projects concern few small hydro power plants (USAID, 2009).

through the CDM framework since such projects seem attractive for investments[60] (BMU, 2008). But, only four Azeri CDM projects are registered up to date due to the current unfavorable investment climate (UNEP Risoe, 2012).

*Bulgaria:* The country ranks in the 39th position regarding attractiveness in RES investments (Ernst & Young, 2012). The investors' interest is not significant since energy from RES is still considerably more expensive compared to energy produced from conventional sources and by traditional technologies – coal, natural gas, nuclear fuel (Bulgarian Government, 2011). Interest in large power storage hydropower plants is limited since they are used for balancing the national electricity network and are operated only in emergency cases (Gaven, 2009). The state fully controls the largest hydropower plants, including 14 power generators holding around 86% of the total hydropower generation in the country (Ganev, 2009). On the contrary, due to unbundling and privatization procedures in the electricity sector, private companies own all small hydropower facilities (Ganev, 2009). Another limitation for competitiveness is the grid connection. SEWRC estimated that connecting renewable energy in the next four to five years, will require EUR 100–120 million (Lefkowitz, 2009).

*Estonia:* Opportunities for investments in RES by foreign investors exist, but are not so attractive compared to other countries (Ernst & Young, 2012). It is positive that the country offers attractive investment opportunities due to low taxes and stable Government. Investors benefit from a 0% tax on reinvested earnings (i.e., investors only pay tax on dividends); equal treatment of domestic and foreign investment; and the flat rate of income tax of 21% regardless of income level (Davies S. and Holmes I., 2011). But the energy market is so small, that deployment of a few 100 MW of wind onto the system equates to a significant market share. There were other significant untapped RES opportunities – particularly biomethane from the farming sector which the Estonia's Renewable Energy Association, estimated at a potential for 300–400 GWh (Davies S. and Holmes I., 2011). Support is already available in the form of grants, but biomethane faces significant barriers. Technology costs are high compared to this financial support and grid reinforcement in rural areas is required for supporting these investments. Another obstacle for investors is the lack of incentive to shift away from oil shale (Davies S. and Holmes I., 2011).

*Moldova:* The attractiveness for RES was very low in 2012 – not included within the 40 most competitive countries of the world (Ernst & Young, 2012). Investors have found the Moldovan market not attractive probably because both the RES and bio-fuels sectors are in early stage and effective support schemes need to be enacted in order to stimulate their growth (Republic of Moldova, 2012; UNECE, 2009).

*Romania:* The country ranks in the 13th position regarding attractiveness in RES investments (Ernst & Young, 2012). Investments in RES facilities are very intense and with a rapid growth rate. 2008 was considered as the year for boosting Romanian wind energy because of the acquisition by the Czech energy company, CEZ, of a wind project located in Dobrogea with total capacity of 600 MW. Apart from wind energy investments (1.4 MW worth of EUR 120 Million) there are investments in small hydroelectric

---

60. the web-site of 'East-Invest provides information of indicative CDM projects (http://www.east-invest.eu/en/Investment-Promotion/Azerbaijan-2/AZ-alternative-energy).

(100 MW, worth of EUR 150 Million), EUR 75 Million for solar collectors, EUR 7.5 Million for PV, EUR 240 Million for heating biomass energy, EUR 280 Million for electricity biomass energy and EUR 15 Million for geothermal energy (Patlitzianas, Karagounis, 2011). Estimated investments in RES till 2020 are at EUR 18.2 Billion (InvestEast, 2010).

The rather low development costs (compared to other EU markets), the higher number of green certificates granted for certain technologies, Structural Funds[61] the price band are encouraging potential investors (Pilifilp, 2012).

*Russia:* There are opportunities for foreign investors but not so attractive compared to other countries (Ernst & Young, 2012). The reasons are: (i) low energy prices compared to international ones, specifically for heat and natural gas, not providing incentives to save energy or returns to attract new investment (Pacific Northwest – National Laboratory, 2012; OECD/IEA, 2005); (ii) delay in the setting of the premium tariff for RES-E and furthermore, it did not ensure: the absence of market risks for investors (since it was linked to the market prices for energy) (UNECE, 2010); a return on investment sufficiently attractive for private investors (UNECE, 2010); (iii) absence of operational dedicated credit lines provided by national funds (UNECE, 2010). The only available financing mechanisms are ordinary commercial loans for short- to mid-term time periods and to relatively high interest rates (UNECE, 2010). (iv) delay in the complete privatization and liberalization process in the energy sector (UNECE, 2010); and (v) strong state influences in the strategic decisions of market operators that limit often the foreign investors to a role of minority shareholders, with no influence on the corporate governance of the companies (UNECE, 2010).

*Serbia:* The country is not included in the 40 most attractive countries for RES. The level of foreign investments for RES is low, although there is interest from foreign investors, which offers knowledge transfer and 'green field' projects (Bogunovic A. and Bogdanov N., 2009). The size of the introduced incentives, the administrative procedures for the construction of these facilities, and the absence of long-term contracts discouraged investors (Bogunovic A. and Bogdanov N., 2009).

There is limited availability of modern technologies for RES. The companies that offer such technologies are not located in Serbia, so products are imported from abroad, which influence the higher price of these installations (Bogunovic A. and Bogdanov N., 2009).

Current prices for energy are not high enough to cover all production costs for the energy producers and much more to proceed with investments in modernization and increasing capacities (Radosavljevic Goran and Djokovic Vuk, 2007). In 2010, the

---

61. The financing of projects in the fields of RES from structural funds is carried out within the Sectoral Operational Programme 'Increase of Economic Competitiveness' (SOP IEC) - Axis 4 'Increasing energy efficiency and security of supply, in the context of combating climate change (*see* http://oie.minind.ro/). The maximum value of the non-refundable support which can be granted for a project as percentage of the eligible expenses is the following: i) for small enterprises and micro-enterprises: 70%, except for projects located in the Bucharest - Ilfov region where the maximum value is 60%; ii) for medium enterprises: 60%, except for projects located in the Bucharest - Ilfov region where the maximum value is 50%; iii) for large enterprises: 50%, except for projects located in the Bucharest - Ilfov region where the maximum value is 40%.

electricity price was EUR 45/MWh, while electricity generation costs were estimated at EUR 30-35/MWh (FAO, 2011). Revenues from other tariff components are not considered as sufficient to cover transmission and distribution costs. Because of this price, big industrial companies preferred to purchase electricity from the Electric Power Industry of Serbia (EPS) instead of producing their own electricity i.e., from RES (FAO, 2011).

Supply and demand are not in proper balance due to the absence of a genuine raw materials market for biomass, (Republic of Serbia, 2010). National and regional chambers of commerce need to identify opportunities for supporting market development and stock exchange. If the domestic market is developed then higher biomass production will be promoted, leading to an established certification process and to job growth (Republic of Serbia, 2010).

There are also shortcomings in legislation such as regulation governing RES utilization, plant design, and installation (Radosavljevic Goran and Djokovic Vuk, 2007). Serbia also largely lacks standards (for equipment, RES extraction procedures, etc.) which are used in the EU (Radosavljevic Goran and Djokovic Vuk, 2007). Serbian CDM projects were not yet registered in the UNEP Risoe[62] database before 2010. Now there are six registered CDM projects mainly for RES.

*Turkey:* Attractiveness in RES investments was improved in 2012 compared to the previous year due to the latest implemented Law for RES (Ernst & Young, 2012). From the 29th position in 2011 for all types of RES it scaled up to the 26th position for 2012 (Ernst & Young, 2012). However, FITs are still low compared to EU countries, although multiple tariffs are envisaged by the amended RES Law (Sirin S.M. and Ege A., 2012). In an effort to reduce its 90% dependence depends on imports of fossil fuels, RES started to be promoted (Ernst & Young, 2013). The European Investment Bank committed for EUR 150 Million of loans to finance RES and EE projects as well as exports (Ernst & Young, 2012). The need for huge investment in this sector has attracted international investors (General Electric Co. and Siemens AG). FDI will play a key role in funding wind, solar, hydropower, biomass, and geothermal energy projects (Ernst & Young, 2013).

*Ukraine:* The country ranks in the 29th position regarding attractiveness in RES investments (Ernst & Young, 2012). The reasons are: (i) Production of power from RES has much lower cost compared to other countries (OECD, 2012). The production of one MWh of wind energy costs roughly USD 33 compared with USD 145 in the Czech Republic, and USD 50 in China (OECD, 2012). Similarly, average costs for producing one KWh of bioelectricity are estimated at EUR 0.057, while the current green tariff in Europe for bioelectricity is EUR 0.127/KWh (OECD, 2012).

Forecasts refer to investments of approximately USD 5 billion for RES generation, including solar and wind energy, biomass, and bio-fuel production, in the next five years (Deloitte, 2012). It is indicative that the level of green tariffs for wind power plants are slightly higher than the average level of EUR 90/MWh across Europe,

---

62. http://www.cdmpipeline.org/.

making this RES type very attractive for local and foreign developers from all over the world (Deloitte, 2012; EBRD/GEF, 2011).

There is difficulty in receiving access to credit resources due to the reluctance of banks to provide loans for investments with a payback period of more than one year (UNECE, 2010). No information is available regarding commercial financing for EE and RES, except for several Carbon Funds, active in the market of JI projects (UNECE, 2010). International investors have complained about no clearly stated criteria for admittance to the wholesale electricity market (OECD, 2012). There were reported cases for which approval was revoked for unclear to the investor reasons and granted only after months of negotiations (OECD, 2012).

The profitability of projects is put at risk due to inefficient market structures and incomplete market opening, with regulated prices on the wholesale market and below-market tariffs for end customers (UNECE, 2010). Municipal administrations apply political tariffs to the supply of heat, thus generating debt and loss of profitability (UNECE, 2010).

Grades: Albania – 4, Armenia – 3, Azerbaijan – 3, Bulgaria – 4, Estonia – 3, Moldova – 3, Romania – 7, Russia – 3, Serbia -3, Turkey – 6, Ukraine – 5.

*Equity:* An index is needed to understand the comparable efforts among these 11 countries in promoting RES. Several indicators are found in literature (Elzen den Michel, Hohne Niklas, Vliet van Jasper, 2009; Vaillancourt Kathleen, Waaub Jean-Philippe, 2004). For this paper, calculations were made to reflect the effort per person to contribute in the consumption of energy from RES per year of the time interval from the base year until the target year (Table 8.3).

*Flexibility:* Based on Table 8.2 regarding the availability of rules and influencing mechanisms that the policy mixture has for the support of RES the following grades were assigned.

Grades: Albania – 4, Armenia – 6, Azerbaijan –4, Bulgaria – 7, Estonia – 5, Moldova – 5, Romania – 6, Russia – 5, Serbia – 4, Turkey – 6, Ukraine – 5.

*Stringency:* Two countries have penalties for the case of non-compliance.

*Bulgaria:* In case that end suppliers do not off take the RES-E as they are obliged to do so, will face a penalty ranging from BGN 7,000 (appr. EUR 3,600) up to BGN 20,000 (appr. EUR 10,225 (KPMG, 2010).

*Moldova:* ANRE found small violations for its licensees, but did not impose sanctions (EBRD, 2009).

*Romania*: Under Law no. 220/2008, there were penalties of EUR 70 for each non-delivered by a supplier GC (KPMG, 2010). Under the new Law no. 138/2010 if suppliers or generators do not comply with their quarterly quota of GCs, they are penalized with the maximum GCs sale price (57.389 EUR/GC – 248.30 RON/GC), for each non-acquired GC (CMS, 2013). The money is collected into a guarantee fund (CSM, 2013). If suppliers or generators do not comply with their annual quota of GCs, they will pay a EUR 110 penalty[63] for each non-acquired GC. The money is collected and used for granting incentives for small RES projects.

---

63. http://www.econet-romania.com/en/article/55/lawsregulations-energy-from-renewable-sources-status-quo.html.

Grades: Albania – 4, Armenia – 4, Azerbaijan – 4, Bulgaria – 7, Estonia – 4, Moldova – 3, Romania – 7, Russia – 4, Serbia – 4, Turkey – 4, Ukraine – 4

*Implementation network capacity:* Each national IN was presented in previous session. Their number, the information that they provide are taken into consideration for the evaluation.

*Albania:* The existing IN does not provide the necessary information, while published reports of certain national entities (such as AKBN, ERE) are not updated. The web-sites are not user friendly and the information is not directly accessible. Users need to devote time in searching for the needed information. The webpages of the Ministry of Economy, Trade and Energy for legislation and monitoring reports are 'Under Construction[64,65]'.

*Armenia:* The Armenian banking system is not ready to fund commercial projects for RES due to lack of knowledge in technologies and economics and to lack of capital for large-scale projects. Furthermore, consolidation of a number of banks increases the cost of the project (USAID, 2010; Danish Energy Management/SolarEn LLC, 2009). The lack of skills and expertise of the IN so as to support policy making is also observed in the National Program. There are difficulties in finding official national reports for RES issues.

*Azerbaijan:* Only since 2009 Azerbaijan started to design a RES policy when it became a signatory to the International Renewable Energy Agency (Energy Charter, 2013). The existing implementation network does not provide the necessary information – at least in English. There is very limited number of official reports regarding RES issues. The web-site of the responsible agency is suspended. No official updated reports are available.

*Bulgaria:* Very limited IN, only one entity responsible for RES. The English version of the SEWRC[66] web-site has less information compared to the Bulgarian one. The information and data regarding RES and RES-E that these entities provide in English through their web-sites are limited. Technological infrastructure and particularly the national grid require upgrading as already aforementioned. Bulgaria is still far from achieving specific grid quality and management capacity for promoting RES (Centre for the Study of Democracy, 2011). The availability of information has been evaluated as negative (ECORYS, 2010).

*Estonia:* Estonia has reported that the public opinion towards a sustainable lifestyle has changed during the past decade due to increased public awareness for energy savings and RES (UNFCCC, 2011). Elering is the TSO that pays out the subsidies according to the Electricity Market Act[67] and manages the electricity system in real time. Its web-site has more information compared to other entities for the renewable energy subsidy,[68] reports and analyses[69] and is kept updated.

---

64. http://www.mete.gov.al/error.php?idr = 2&lang = 2.
65. http://www.mete.gov.al/error.php?idr = 535&lang = 2.
66. www.dker.bg.
67. http://elering.ee/renewable-energy-3/.
68. http://elering.ee/renewable-energy-3/.
69. http://elering.ee/reports-and-analyses/.

*Moldova:* The English version of the web-site of ANRE[70] has limited information compared to the Romanian version. Investments in RES facilities are made mainly in biomass and solar for heating production based on foreign assistance[71] and not national resources (SNC, 2009). No feasibility studies of sector potential or impact of the rate design were conducted, making the long-term impact unclear (EBRD, 2009). The availability of information has been evaluated as positive (ECORYS, 2010).

*Romania:* The information and data regarding RES that the IN entities provide in English through their web-sites are limited. The Romanian ANRE[72] has updated web-site for the electricity market and the legislative framework for RES. Technological infrastructure and particularly the national grid require upgrading as already aforementioned (GWEC, 2011). The availability of information has been evaluated as negative since public awareness-raising campaigns have been limited and/or conceived and carried out at low quality levels (ECORYS, 2010).

*Russia:* The existing IN does not provide the necessary information, at least in English. There is limited number of official reports in English, while those in Russian language have general information. The development of analysis work and assessments for supporting the awareness of customers and for providing a quantitative background for the identification of attractive EE and RES projects is restricted due to the partial absence of databases with consumption data on public and residential buildings (UNECE, 2010). Established on line databases are not updated or do not provide the necessary information (UNECE, 2010). There is no entity directly responsible for the exploitation of RES in Russia.

The current IN does not provide the necessary information of how regulations are applied and enforced and how property rights are protected (World Economic Forum, 2013). This situation along with corruption[73] and undue influence affects negatively economic transactions and creates problems for doing business in Russia (World Economic Forum, 2013). Barriers for investments pointed out the existence of general and diffused lack of awareness on the relevance of EE and RES in RF (UNECE, 2010). The absence of operational credit lines among financial institutions indicates a general lack of experience in the national banking sector concerning financing schemes for EE and RES projects (UNECE, 2010).

*Serbia:* AERS has limited information about RES although the web-site is updated. Public awareness and knowledge for specialized issues such as cogeneration and biomass is low (Danon et al., 2012; Embassy of Denmark in Belgrade, 2010; Bogunovic and Bogdanov, 2009). There is lack of information (apart from lack of capital), lack of experience among potential investors, authorized persons in local self-governments, etc. (Danon et al., 2012; Bogunovic and Bogdanov, 2009).

Serbia suffers from underdeveloped institutions, lack of expertise at all levels of government (Republic of Serbia, 2010; Jefferson Institute, 2009). Serbian policy makers

---

70. http://www.anre.md/index.php?vers = 3.
71. http://www.undp.md/presscentre/2012/Biomass_23July/index_rom.shtml.
72. www.anre.ro.
73. According to Transparency International, public officials and civil servants, including the police, are classified as the most corrupt institutions in the country, followed by the education system and parliament (World Economic Forum, 2013).

need to provide and secure training so as to increase the number of energy experts, to enrich curricula of the education system with relevant topics, put in place the necessary regulations, and ensure the appropriate amount of energy prices (Embassy of Denmark in Belgrade, 2010; Jefferson Institute, 2009). Due to the absence of qualified personnel to implement existing legislation, there are delays and poor communication among regulatory institutions, and private and civic actors (Jefferson Institute, 2009).

The statistical methodology of data collection concerning biomass was not in accordance with the systems used in EU countries (Republic of Serbia, 2010). Serbian consumers are not familiarized with the term 'bio-energy' and do not realize advantages of biomass (Republic of Serbia, 2010; Bogunovic and Bogdanov, 2009). There is lack of clear communication about this term, especially between energy companies and environmental organizations, with the latter not believing that all bio-energy options are equally sustainable (Republic of Serbia, 2010; Bogunovic and Bogdanov, 2009).

The creation of definitive guides for the RES licensing process from the IN for encouraging investors and other market actors will improve the situation. Governmental authorities have created 'roadmaps', for explaining the licensing process, the development of projects in geothermal water, small hydropower, wind power, and biomass subsectors (Danon et al., 2012).

Although producers and suppliers of equipment used for RES provide the necessary information on characteristics of such equipment and systems in their web-sites and arrange presentations to interested persons,[74] this does not happen with national relevant web-sites (Republic of Serbia, 2013). The information system needs to provide information to a larger number of interested parties, particularly to natural and legal persons which cannot attend public presentations (Republic of Serbia, 2013). The preparation of such a web-site will be based on existing database of companies dealing with RES issues (Republic of Serbia, 2013). Additionally, market activities – for RES and particularly for biomass – for the stock exchange (price changes, trading volume, selling places, linking participants, possibilities for export) need to be transparent and accessible to all stakeholders and participants through web information or publications (Republic of Serbia, 2010).

*Turkey:* In its web-site the Ministry of Energy and Natural Resources[75] has a session for publications which is not updated. Its session about 'Statistics' is under construction. The documents on the 'Energy-Environment-Climate Change'[76] session are in the Turkish language only. The Ministry of Environment and Urban Development[77] does not have an English version for its web-site. The EMRA[78] has limited information about RES in its web-site. It is indicative that EBRD will continue to seek opportunities to provide technical assistance to the government under the framework

---

74. Serbian Chamber of Commerce and Industry, Serbian Chamber of Engineers, R&D organizations and professional organizations and association (Republic of Serbia, 2013).
75. http://www.enerji.gov.tr/index.php?dil = en&sf = webpages&b = yayinlar_raporlar_EN&bn = 5 50&hn = &id = 40721.
76. http://www.enerji.gov.tr/index.php?dil = en&sf = webpages&b = enerji_cevre_iklim_EN&bn = 218&hn = &id = 40720.
77. http://www.csb.gov.tr/turkce/index.php.
78. http://www.emra.org.tr/.

of the Sustainable Energy Action Plan aiming to strengthen the institutional capacity and approaches for ensuring timely compliance with the European Climate Change package (EBRD, 2012).

*Ukraine:* The IN is not adequate since technical capabilities and professional skills might be available, but there is no expertise available for project financing and profitability evaluation of EE and RES projects (UNECE, 2010). Another disadvantage of the IN is the strong lack of transparency regarding the amount of financing and the allocation criteria for the State Fund for Energy Conservation (UNECE, 2010). There is no dedicated agency or institution for the development of RES (UNECE, 2010). This task was assigned to the SAEEEC, but this entity has institutional priorities regarding the promotion of EE (UNECE, 2010).

Grades: Albania – 2, Armenia – 3, Azerbaijan – 2, Bulgaria – 4, Estonia – 8, Moldova – 3, Romania – 5, Russia – 2, Serbia – 2, Turkey – 4, Ukraine – 2.

*Administrative feasibility:* Emphasis is given to the administrative complexity that the policy mixture imposes.

*Albania:* Regarding investments in RES the Government of Albania tried to support the development incentives of the private sector by improving the legal framework, simplifying the administrative procedure for investors and guaranteeing a transparent and non-discriminated process for the interested subjects (AKBN, 2010b). However, the need of an improved regulatory framework remains (EBRD, 2012).

*Armenia:* The pertinent governmental authorities for EE and RES policy issues have delayed in implementing the necessary legislative framework (Armenia Renewable Resources and Energy Efficiency Fund, 2008; Energy Charter Protocol on Energy Efficiency and Related Environmental Aspects PEEREA, 2005). They did not work, design and implement the necessary regulations or programs and they did not proceed with the establishment of the necessary institutions (Armenia Renewable Resources and Energy Efficiency Fund, 2008). The currently implemented policy mixture performs low because: (i) it is time-consuming in obtaining the necessary permits and licenses for SHPP and wind power generation (Danish Energy Management (A/S), 2011); (ii) there is insufficient coordination between different authorities in obtaining permits; and (iii) there are problems related to transparency in procedures, long lead-times and high costs involved in obtaining permits or licenses (Danish Energy Management (A/S), 2011).

*Azerbaijan:* There were delays with the implementation of the policy instruments and the preparation and participation of target groups in proper time. There is a need for developing laws and secondary legislation on EE and RES (Energy Charter, 2013).

*Bulgaria:* Administrative procedures are complex and lengthy since investors need various certificates, permissions, and licenses issued by different authorities (for planning permission, permission for change of land use, positive resolution on the environment impact assessment of the project, water permit (for hydropower projects), construction permit, generation license, etc.) (KPMG, 2010). SEWRC has most of the main responsibilities for the implementation of RES projects which creates administrative burden (Ganev, 2009). Insufficient support for RES is attributed to the administrative incapability to formulate and implement policies (Centre for the Study of Democracy, 2011). Administrative delays are frequently observed in the process of connecting RES to the grid (Centre for the Study of Democracy, 2011). Administrative deficiencies are linked with corruption regarding public procurement and permit issuing procedures (Centre for the Study of Democracy, 2011).

*Estonia:* National measures concerning green economy are fragmented, unsystematic and sometimes contradict the measures applied by government for supporting the current economic structure (Valdur, 2012). Currently, there is no comprehensive national strategy dedicated to the promotion of green economy and/or low-carbon economy (Valdur, 2012). A high number of authorities are involved in permitting, while there is lack of experience (ECORYS, 2010). The civil servants dealing with permitting procedures are not always familiar with RES. So, confusion, delays, or unmotivated denials of authorizations emerge (ECORYS, 2010).

Regarding awareness-raising programs applied in the recent years, there is no information on how these programs are monitored (UNFCCC, 2011). The lack of competent institutions for climate change issues shows that all the work is concentrated at the relevant Ministries increasing the administrative burden.

Promotion of RES is linked also with the opening of the electricity market. According to the agreement achieved, during the European Union accession negotiation, the opening up to 35% of the electricity market was scheduled for the end of 2008 (Treaty, 2003). Estonia was granted a transition period until 31 December 2015 regarding also the implementation of the EU Directive on reduction of emissions into air from large combustion plants (2001/80/EC) like the oil shale-fired power plants. Estonia was granted this transition period so as to manage successfully the Directive, which means that the Ministry of Economic Affairs and Communications, Elering, and other ministries need to coordinate the implementation of the policy instrument (Laur, Soosar, 2002).

*Moldova:* The institutional and administrative capacity of the country for energy and climate change issues requires strengthening, particularly for strategic planning, implementation and enforcement.[79] For the period 2003-2006, ANRE suffered by limited ability to function because the government did not appoint a third regulator (EBRD, 2009). The preparation of a National Action Plan for RES is delayed, while no tendering mechanisms or specific incentive programs are in place from the Government. The adoption of a methodology on RES and bio-fuel (not separated by type) and draft supply contracts for them was in time (EBRD, 2009). ANRE is able to issue secondary legislation on tariffs, licenses; dispute resolution; set and revise tariffs; issue, suspend, or revoke licenses, impose fines for infractions, and may issue orders. By law, there is no overlap in authority and the Ministry has no direct authority over ANRE.

*Romania:* Administrative burden is significant due to an excessive number of authorities involved in licensing procedures (Pirlogea, 2011; ECORYS, 2010). The application for obtaining a permit contains documents and documentation that fulfill all requirements of economic, financial, technical and professional nature set for the electricity generation capacity and the specific activities (GWEC, 2011). ANRE issues

---

79. European Neighbourhood and Partnership Instrument, Republic of Moldova – Country Strategy Paper 2007–2013. Available at: http://ec.europa.eu/world/enp/pdf/country/enpi_csp_moldova_en.pdf.

its decision after a period of 30 days, maximum, from the date on which it has established that the submitted documentation is complete and that the applicant has paid all taxes and provided proof of payment (GWEC, 2011). There is lack of experience as in the Estonian case (ECORYS, 2010).

*Russia:* There were no action or implementation plans developed for the previous 'Energy Strategy until 2020' (UNECE, 2010). There was a lack of: (i) a clear policy framework for the substantial support of EE and RES; (ii) clear responsibilities for policy implementation and monitoring of progress and this clearly constituted a significant legal and institutional barrier (UNECE, 2010). Particularly for RES, there is still no defined regulatory framework in place or under development (UNECE, 2010). There is administrative burden – caused by inefficient or unclear bureaucratic processes – which leads to increased transaction costs within the economy, from opening a business to customs procedures or accessing utilities (World Economic Forum, 2013). It is indicative that obtaining a new electricity connection requires 10 procedures, takes 281 days, and costs 1.852% of per capita income to complete (World Bank, 2012).

*Serbia:* For RES facilities the analysis of existing procedures for obtaining licenses, permits and approvals showed that they (Republic of Serbia, 2013; Bogunovic and Bogdanov, 2009): (i) are long and demanding since permits need to be obtained in a specific order; (ii) are costly; require knowledge of all documentation and deadlines (by-law documents); (iii) require familiarization with a great number of laws and by-laws defining competencies for obtaining of licenses, permits, and approvals. Additionally, issues related to the ownership of the land, legal regulations linked to giving of concessions and locations are problematic (Bogunovic and Bogdanov, 2009). All these administrative procedures for the construction of RES facilities discourage investors (Bogunovic and Bogdanov, 2009).

Administrative burden is increased due to the existence of differences in terms between laws and regulations for biomass potential and its utilization (Republic of Serbia, 2013; Republic of Serbia, 2010). The administrative inertia in combination with the lack of knowledge about the possibility of biomass utilization contributed to the low development level in this area (Republic of Serbia, 2010). Low-ranking authorities, policymakers, and relevant institutions are not always aware of the advantages and disadvantages of bio-energy, which leads to a cautious approach in licensing procedures and causes delays (Republic of Serbia, 2010). In addition, only a relatively small number of bio-energy projects have been realized to date in the country, so there is not much experience with licensing in this area (Republic of Serbia, 2010).

The licensing procedure must be clearly defined (even for the responsibilities of institutions at different levels), adjusted specifically for certain RES types,[80] simplified without omitting important elements regarding the safety of the facilities, energy and ecological requirements, etc. (Republic of Serbia, 2010). Investors need to know which level of government is responsible for issuing permits (state or municipal) and which institutions are responsible for issuing permits (Republic of Serbia, 2010). The process must be monitored so as to improve it (Republic of Serbia, 2010).

---

80. There are examples where procedures for the other types of facilities were used for biomass (Republic of Serbia, 2010).

There is not an effective regulatory framework designed to increase EE and advance the use of RES (Jefferson Institute, 2009). The necessary sub-laws and regulations are not introduced so as to enable the full implementation of the already set into force laws (Jefferson Institute, 2009). The Revised FITs, which were due by the end of 2011, were not adopted in 2012 (European Commission, 2012). The biggest obstacle for the exploitation of RES is the administrative procedures for issuing construction permits, licensing and network connections (European Commission, 2012). This situation requires further efforts for strengthening administrative capacity and creating a regulatory environment that encourages the increased use of RES in all sectors (European Commission, 2012). Serbia has not yet adopted the planned framework law on rational use of energy.

*Turkey:* EMRA is the main responsible entity, which increases administrative burden. With the 2005 Law there were problems about administrative hurdles, increasing wholesale prices and growing local opposition particularly against small hydroelectric plants (Sirin S.M. and Ege A., 2012). There are still administrative and time-consuming barriers for foreign investors, which require the establishment of an institutional entity, with high level of coordination and cooperation within and between institutions, agencies, institutes, and other stakeholders (Erdogdu, 2009). There are important uncertainties regarding the Renewable Energy Support Mechanism (PWC, 2012). None of the participants in this Mechanism could receive the premium in 2012 due to a disagreement between investors and the Energy Ministry over the term 'locally manufactured equipment' (PWC, 2012).

*Ukraine:* Further strengthening of administrative capacity at all levels of the country and coordination between the authorities requires particular attention (CEC, 2009). Furthermore, the FITs for RES is in place since 2008, but the coefficients for the calculation of prices were published only in spring 2009 (UNECE, 2010). This policy instrument still needs time for realizing if it is fully effective and make the necessary modifications. The legal framework for the energy sector in general and particularly that referring to EE and RES is complex, partly fragmented, and partly outdated (UNECE, 2010).

The Ukrainian 'Energy Strategy until 2030' contains targets for EE and RES, but these targets are rather ambitious (50% reduction of energy intensity and share of 19% of RES in primary energy supply) and, in absence of an implementation plan, it is not clear how these targets will be achieved (UNECE, 2010). Ukraine does not regularly review its RES strategy targets as these are defined in the 'Energy Strategy until 2030' (OECD, 2012).

Administrative procedures are a significant barrier to RES growth (OECD, 2012). Investors need to deal with six or more government agencies and institutions so as to comply with permitting procedures (OECD, 2012). This process requires time, increases the risk levels for businesses and increase the possibility for corruption (OECD, 2012).

Administrative issues are related to gaps in the legislation. NERC provides green tariffs to projects that have fulfilled the required documentation (EBRD/GEF, 2011). NERC has found some difficulties in defining those tariffs when the involved input fuel was biomass, since in the Law of Electricity there is no coefficient defined for electricity produced from the biomass of animal origin (biogas from animal origin) (EBRD/GEF, 2011).

Grades: Albania – 4, Armenia – 4, Azerbaijan – 2, Bulgaria – 4, Estonia -5, Moldova – 4, Romania – 5, Russia – 3, Serbia – 2, Turkey – 4, Ukraine – 4.

*Financial feasibility:* Emphasis is given if the government can afford the financial cost of its policy mixture and what financial resources are used.

*Albania:* Limited financial resources for the policy mixture and existence of donors are the two elements that characterize the performance under this sub-criterion. EBRD through its updated 'Strategy for Albania' will: (i) assist the Albanian Power Corporation so as to improve its financially viability and stability, by providing long-term loans for the safety, maintenance, and upgrading of its generation utilities as well as for a possible restructuring of its operations (EBRD, 2012); (ii) continue to implement the WeBSEFF/WeBSEDFF frameworks for the financing of RES and EE projects by extending support to private concessionaires and companies implementing energy savings (EBRD, 2012). Bilateral donors (Sweden, Italy) are active and important partners for Albania by providing grant funds or soft loan financing, and assisting sectors with lower cost-recoverability (EBRD, 2012).

*Armenia:* The fact that there are low tariffs for supporting different RES types and no policy instruments (taxes, fees, levies, or heavy penalties) that could create revenues show that national financial resources are limited. However, the country tried and has established cooperation with international entities that support RES projects (i.e., European Bank for Reconstruction and Development (EBRD), International Finance Corporation (IFC) (member of the World Bank Group), World Bank).

*Azerbaijan:* There is a EUR 14 million budget support program aiming to assist the Azeri Ministry of Energy in developing a legislative and regulatory framework for RES and EE (EC, 2012). Simultaneously, EBRD intends to continue supporting Azerbaijan's infrastructure development with emphasis on the power sector and energy saving investments (EBRD, 2010).

*Estonia:* Estonia has a major financing problem, since the general economic conditions are not conducive and the government has little funding of its own for EE or RES investments (Egger C. et al., 2012). There are revenues from the sale of emission rights (AAUs) under the Kyoto Protocol ang through the GIS (starting mainly from 2013) (MOIEEP, 2011). Due to the availability of funds, experts see significant progress in financial instruments (Egger C. et al., 2012).

*Moldova:* Limited resources, based on donors like other countries. No available information.

*Romania:* Limited resources, the aforementioned fund assists in the financial feasibility of the policy mixture. EU EU-funds for alternative energy projects have been part of the 2007–2013 programs for Structural Funds (Romanian Business Digest, 2013).

*Russia:* The absence of any tariff components to cover environmental externalities leads to diminished awareness of the value of natural resources on behalf of the Russian customers and implies the absence of a potential financial mechanism for EE and RES projects (UNECE, 2010). The latter is actually happening. The implementation of RES projects needs State-sponsored support system, but the current legislation fails to provide any financial incentives for RES generation (Kopylov A., 2010). The development of relevant policy instruments in support of RES remains a necessity.[81] Such institutions are: EBRD, IFC (member of the World Bank Group); Developing Renewable Energy in Russia;[82] Nordic Environment Finance Corporation (NEFCO)[83] (UNECE, 2010).

*Serbia:* The Government has the following options (Republic of Serbia, 2013): (i) international sources of financing: the German Development Bank (KfW), Fund Green for Growth, International Financial Corporation /IFC, EBRD. (ii) local sources of financing.

*Turkey:* EBRD will provide investments to private sector investors for RES and efficient power production and will support the ongoing reforms of the energy sector (EBRD, 2012). EBRD provided one loan to a large wind farm project and loans to partner banks under two supportive frameworks, one for small scale EE projects and the other for mid-sized RES projects (EBRD, 2012). Two more banks assisting towards these efforts are: EIB and Islam Development Bank. AFD contributes to investment in RES and promoting EE in industrial enterprises, particularly SMEs and in housing (ADB, 2012) AFD and Proparco, its private sector financing arm, support RES development via partner banks (i.e., TSKB, Garanti Bankası, İş Bankası and Aklease), and with direct investments also (AFD, 2012).

*Ukraine:* The governmental sources of financing are the State (national) Budget and local budgets at all levels (UNECE, 2010). So far, no funds are assigned for the development of RES (UNECE, 2010). Budgets/ funds for energy saving measures/ investments have been inadequate (Ministry of Strategy and Finance, the Republic of Korea, 2010). The financial resources for promoting RES are from: (i) International sources (BSTDB: EBRD, Neighbourhood Investment Facility (NIF) (under the European Neighbourhood and Partnership Instrument) and (ii) National sources: GIS revenues (World Bank, 2006).

Grades: Albania – 3, Armenia – 3, Azerbaijan – 3, Bulgaria – 4, Estonia – 5, Moldova – 3, Romania – 5, Russia – 3, Serbia – 4, Turkey – 4, Ukraine – 4.

---

81. *See* for example the article of Chairman of the State Duma Committee on International Affairs K.Kosachev 'Alternative sources of energy: Russia and the world experience' in 'Rossiyskaya gazeta' http://www.rg.ru/2011/06/08/kosachev-poln.html.
82. http://www.ifc.org/wps/wcm/connect/region__ext_content/regions/europe+middle+east+and+north+africa/ifc+in+europe+and+central+asia/countries/developing+renewable+energy+in+russia and http://www.ifc.org/wps/wcm/connect/RegProjects_Ext_Content/ifc_external_corporate_site/home-rrep.
83. http://www.nefco.org/introduction/this_is_nefco.

Table 8.4  Results of the AMS Method (calculations by author)

| Criteria | Countries | | | | | | | | | | |
|---|---|---|---|---|---|---|---|---|---|---|---|
| | Albania | Armenia | Azerbaijan | Bulgaria | Estonia | Moldova | Romania | Russia | Serbia | Turkey | Ukraine |
| Direct contribution to GHG emission reductions (0,833) | 79.337 | 0.000 | 7.240 | 40.011 | 83.300 | 21.797 | 47.633 | 35.629 | 1.486 | 6.402 | 17.605 |
| Indirect environmental effects (0,167) | 15.944 | 15.812 | 13.870 | 16.674 | 14.951 | 16.700 | 8.965 | 7.032 | 11.813 | 2.637 | 0.000 |
| **Environmental performance (0,168) – A** | 95.281 | 15.812 | 21.110 | 56.685 | 98.251 | 38.497 | 56.598 | 42.661 | 13.299 | 9.039 | 17.605 |
| Cost efficiency (0,473) | 31.407 | 0.000 | 0.000 | 47.300 | 47.300 | 0.000 | 31.407 | 0.000 | 0.000 | 0.000 | 23.650 |
| Dynamic cost efficiency (0,182) | 1.224 | 1.224 | 0.766 | 3.047 | 4.826 | 1.224 | 1.908 | 0.766 | 1.224 | 0.766 | 1.224 |
| Competitiveness (0,085) | 0.608 | 0.390 | 0.390 | 0.608 | 0.390 | 0.390 | 2.436 | 0.390 | 0.390 | 1.538 | 0.971 |
| Equity (0,175) | 6.076 | 0.756 | 0.000 | 4.168 | 17.500 | 0.986 | 4.002 | 2.253 | 0.187 | 1.044 | 0.115 |
| Flexibility (0,050) | 0.238 | 0.603 | 0.238 | 0.955 | 0.380 | 0.380 | 0.603 | 0.380 | 0.238 | 0.603 | 0.380 |
| Stringency for non-compliance (0,034) | 0.204 | 0.204 | 0.204 | 0.818 | 0.204 | 0.131 | 0.818 | 0.204 | 0.204 | 0.204 | 0.204 |
| **Political acceptability (0,738) – B** | 39.757 | 3.177 | 1.599 | 56.896 | 70.601 | 3.112 | 41.175 | 3.994 | 2.244 | 4.154 | 26.544 |

## Chapter 8: Evaluation of Implemented Policy Mixtures for RES §8.05

| Criteria | Countries | | | | | | | | | | | |
|---|---|---|---|---|---|---|---|---|---|---|---|---|
| | Albania | Armenia | Azerbaijan | Bulgaria | Estonia | Moldova | Romania | Russia | Serbia | Turkey | Ukraine |
| Implementation network capacity (0,309) | 5.461 | 1.498 | 0.937 | 2.334 | 14.823 | 1.498 | 3.727 | 0.937 | 0.937 | 2.334 | 0.937 |
| Administrative feasibility (0,581) | 0.937 | 5.461 | 2.193 | 5.461 | 8.721 | 5.461 | 8.721 | 3.504 | 2.193 | 5.461 | 5.461 |
| Financial feasibility (0,110) | 0.679 | 0.679 | 0.689 | 1.058 | 1.688 | 0.679 | 1.688 | 0.679 | 1.058 | 1.058 | 1.058 |
| **Feasibility of implementation (0,094) – C** | 7.077 | 7.637 | 3.809 | 8.853 | 25.232 | 7.637 | 14.136 | 5.120 | 4.188 | 8.853 | 7.456 |
| Total (A+B+C) | 46.013 | 5.719 | 5.084 | 52.344 | 70.981 | 9.482 | 41.224 | 10.596 | 4.284 | 5.417 | 23.248 |

## [A] Results

The Results are presented in Table 8.3. Estonia is the country whose policy mixture for RES is more effective compared to the other 10 countries. The Serbian policy mixture for RES is the less effective.

## §8.06 CONCLUSIONS

The combination of the two research tools, LEAP and AMS, allowed the quantitative comparison of the performance that 11 national policy mixtures have. The policy mixtures of 4 countries, Estonia, Bulgaria, Albania, and Romania have the higher total grades. Estonia and Albania will probably surpass the set national target, while Bulgaria and Romania will achieve it.

The common characteristics of all four policy mixtures that contribute to their effective performance are: Direct contribution to GHG emission reductions, cost efficiency, equity, and implementation network capacity.

The usage of RES not only in the Energy sector, but also in sectors such as households and transport over fossil fuels led to GHG emission reductions. The combination of policy instruments was more cost efficient for the target groups in their efforts to use more RES and to emit less. Because of the implementation of policy instruments that concerned a wider range of national economic sectors, equity was higher allowing a fairer burden sharing. The more capable an IN is in disseminating information and assisting target groups to understand their options so as to comply the more possible is to facilitate the implementation of the policy mixture.

For two of these countries, Bulgaria and Romania, another characteristic is also important, the pair flexibility and stringency for non-compliance. The performance of their policy mixtures is higher for both sub-criteria compared to those of the other countries. This implies that a policy mixture can have a variety of option for the target groups, but when they fail they have to face consequences.

Based on the aforementioned comments, policy and decision makers are able to understand the strengths and the weaknesses of an implemented policy mixture and to proceed with the necessary modifications so as to increase its effectiveness.

## REFERENCES

AFD, 2012. http://www.afd.fr/webdav/shared/PORTAILS/PUBLICATIONS/PLAQUET TES/AFD-turquie-va.pdf.

AgriPolicy, Enlargement Network for Agripolicy Analysis, 2009. Analysis of renewable energy and its impact on rural development in Estonia. http://www.euroquality files.net/AgriPolicy/Report%202.2/AgriPolicy%20WP2D2%20Estonia%20Final.pdf19.

AgriPolicy, 2009b. Analysis of Renewable Energy and its impact on rural development in Albania. AgriPolicy, Enlargement Network for Agripolicy Analysis. Available

at: http://www.euroqualityfiles.net/AgriPolicy/Report%202.2/Agripolicy%20 WP2%20D2%20Albania%20Final.pdf.
AKBN, 2010. Invest in Albanian Natural Resources – Renewable Energies in Albania. Available at: http://www.akbn.gov.al/images/pdf/publikime/E._Rinovueshme.pdf.
AKBN, 2010b. Invest in Albanian Natural Resources – Renewable Energies in Albania – Hydroenergetic potential. Available at: http://www.akbn.gov.al/images/pdf/publikime/Hydroenergy.pdf.
Armenia Renewable Resources and Energy Efficiency Fund, 2008. The Other Renewable Resource: The Potential for Improving Energy Efficiency in Armenia – Report to the World Bank. Available at: http://r2e2.am/wp-content/uploads/2012/07/The-Potential-for-Improving-Energy-Efficiency-in-Armenia.pdf.
BP, 2012. BP Energy Outlook 2030. London, January 2012. At: http://www.bp.com/liveassets/bp_internet/globalbp/STAGING/global_assets/downloads/O/2012_2030_energy_outlook_booklet.pdf.
Bogunovic Aleksandar & Bogdanov Natalija, 2009. Analysis of Renewable Energy and its Impacts on Rural Development in Serbia. November 2009 – AgriPolicy Enlargement Network for Agripolicy analysis. Available at: http://www.euroqualityfiles.net/AgriPolicy/Report%202.2/AgriPolicy%20WP2D2%20Serbia%20Final.pdf.
Boute Anatole, 2012. Promoting renewable energy through capacity markets: An analysis of the Russian support scheme. Energy Policy 46, pp. 68–77.
Boute Anatole, 2011. A Comparative Analysis of the European and Russian Support Schemes for Renewable Energy: Return on EU Experience for Russia.
Bulgarian Government, 2011. *Energy Strategy of the Republic of Bulgaria till 2020, For reliable, efficient and cleaner energy*, at: http://www.emi-bg.com/userfiles/109.pdf.
Commission of the European Communities, 2008. Commission Staff Working Document – The support of electricity from RES. *Accompanying document to the Proposal for a Directive of the European Parliament and of the council on the promotion of the use of energy from renewable sources.* Brussels, 23 January 2008, SEC(2008) 57, {COM(2008) 19 final}. At: http://ec.europa.eu/energy/climate_actions/doc/2008_res_working_document_en.pdf.
CMS, 2013. Renewables Support Mechanisms Across Europe, A comparative study. At: http://www.cms-dsb.com/Hubbard.FileSystem/files/Publication/66d448bf-8611-4409-bb4f-4ccee4bbbcf9/Presentation/PublicationAttachment/d2078c0b-4df4-45f2-9e72-359f37f345fb/CMS_Renewable_Energy_Guide_April_2013_b.pdf.
Davies Susan & Holmes Ingrid, 2011. European Perspectives on the Challenges of Financing Low Carbon Investment: Estonia – September 2011. Available at: http://www.e3g.org/docs/E3G_European_Perspectives_on_the_Challenges_of_Financing_Low_Carbon_Investment_Estonia.pdf.
Danish Energy Management (A/S), 2011. Renewable Energy Roadmap for Armenia-Task 4 Report, Submitted to Armenia Renewable Resources and Energy

Efficiency Fund (R2E2), May 2011. Available at: http://r2e2.am/wp-content/uploads/2012/07/Renewable-Energy-Roadmap-for-Armenia.pdf and http://www.arlis.am/DocumentView.aspx?DocID=40316.

Danish Energy Management/SolarEn LLC, 2009. Task 3 – Market Penetration Plan Development and Sustainability Recommendations: Final Report 'Assessment of PV Industry Development Potential In Armenia'. Available at: http://www.r2e2.am/documents/english/solar_task3_eng.pdf and http://r2e2.am/wp-content/uploads/Task-3-Solar-PV-Final_Report_eng.pdf.

Danon Gradimir, Furtula Mladen & Mandic Marija, 2012. Possibilities of implementation of CHP (combined heat and power) in the wood industry in Serbia. Energy, Vol. 48, Issue 1, pp. 169–176.

Deloitte, 2012. Industry overview – Renewable energy in Ukraine. Available at: http://investukraine.com/wp-content/uploads/2012/06/Renewable-energy-in-Ukraine_230_230_WWW.pdf.

den Elzen Michel, Hohne Niklas, van Vliet Jasper, 2009. Analysing comparable greenhouse gas mitigation efforts for Annex I countries. Energy Policy 37, pp. 4114–4131.

EBRD, 2009a. Estonia, Country Profile. http://ws223.myloadspring.com/sites/renew/Shared%20Documents/2009%20Country%20Profiles/estonia.pdf.

EBRD, 2009. Strategy for Armenia – As approved by the Board of Directors at its meeting on 21 April 2009. At: http://www.google.gr/url?sa=t&rct=j&q=investments%20in%20armenia%20from%20ebrd%20for%20energy&source=web&cd=40&cad=rja&ved=0CG4QFjAJOB4&url=http%3A%2F%2Fwww.enpi-info.eu%2Flibrary%2Fsites%2Fdefault%2Ffiles%2Fattachments%2Farmenia.pdf&ei=pNzZUaiKGMSEO-XNgZAC&usg=AFQjCNGg1DqkqM8LYmntGNqHxL1y1BC_Vw&bvm=bv.48705608,d.d2k.

EBRD, 2012. Strategy for Turkey. As approved by the Board of Directors at its meeting on 17 April 2012. At: http://www.ebrd.com/downloads/country/strategy/turkey.pdf.

EBRD, 2012. Strategy for Albania – As approved by the Board of Directors at its meeting on 11 December 2012. Available at: http://www.ebrd.com/downloads/country/strategy/albania.pdf.

EBRD, 2009. http://www.ebrd.com/downloads/legal/irc/countries/moldova.pdf.

EBRD/GEF, 2011. Assistance to the National Energy Regulatory Commission of Ukraine: Regulatory Support Programme – TCS ID: 29084, Subtask I: Full implementation of the existing Green Tariff Law and Methodology. At: http://www.uself.com.ua/fileadmin/documents/Task_1_Final_Report_eng.pdf.

EC, 2006. Available at: http://ec.europa.eu/energy/res/legislation/doc/electricity/member_states/2006/estonia_en.pdf.

ECORYS, 2010. Assessment of non-cost barriers to renewable energy growth in EU Member States – AEON, DG TREN No. TREN/D1/48 – 2008, Final report. At: http://ec.europa.eu/energy/renewables/studies/doc/renewables/2010_non_cost_barriers.pdf.

Embassy of Denmark in Belgrade, 2010. Sector Analysis – Sector: Environment. At: http://serbien.um.dk/en/ ~ /media/Serbien/Documents/TCreports2/Environm entSerbia.ashx.

Energy Charter Protocol on Energy Efficiency and Related Environmental Aspects PEEREA, 2005. Armenia, Regular Review 2005 – Part I: Trends in energy and energy efficiency policies, instruments and actors. Available at: http://www.en charter.org/fileadmin/user_upload/document/Energy_Efficiency_-_Armenia_-_2005_-_ENG.pdf.

Energy Charter, 2013. In-Depth Review of the Energy Efficiency Policy of Azerbaijan. At: http://www.encharter.org/fileadmin/user_upload/Publications/Azerbaijan_EE_2013_ENG.pdf.

Ernst & Young, 2012. Renewable energy country attractiveness indices. Issue 34, August 2012. At: http://www.ey.com/Publication/vwLUAssets/Renewable_energy_country_attractiveness_indices_-_August_2012/$FILE/Renewable_energy_country_attractiveness_indices_Aug_2012.pdf.

Energy Community, 2012. Decision of the Ministerial Council of the Energy Community, D/2012/04/MC-EnC: Decision on the implementation of Directive 2009/28/EC and amending Article 20 of the Energy Community Treaty, Ref.: 10TH MC/18/10/2012-Annex18/09.07.2012. Available at: http://www.energy-com munity.org/pls/portal/docs/1766219.PDF.

Erdogdu Erkan, 2009. On the Wind Energy in Turkey, Energy Market Regulatory Authority, Republic of Turkey. At: http://mpra.ub.uni-muenchen.de/19096/1/on_the_wind_energy_in_turkey_word.pdf.

EREC, 2011. Mapping Renewable Energy Pathways towards 2020 – EU Roadmap. Available at: http://www.repap2020.eu/fileadmin/user_upload/Roadmaps/EREC-roadmap-V4_final.pdf.

Ernst & Young, 2008. *Renewable energy country attractiveness indices.* Not available on line, available through contacting Ernst & Young at: http://www.ey.com/UK/en/Industries/Cleantech/Renewable-Energy-Country-Attractiveness-Index – Archive.

Ernst & Young, 2012a. Renewable energy country attractiveness indices. Issue 34, August 2012. Available at: http://www.ey.com/Publication/vwLUAssets/Renewable_energy_country_attractiveness_indices_-_August_2012/$FILE/Renewable_energy_country_attractiveness_indices_Aug_2012.pdf.

Ernst & Yound, 2012b. Growing Beyond – Positioned for growth, Ernst & Young's 2012 attractiveness survey – Russia. Available at: http://www.ey.com/Publication/vwLUAssets/Positioned_for_growth/$FILE/Positioned_for_growth.pdf.

Ernst & Young, 2013. Growing beyond – Ernst & Young's attractiveness survey, Turkey 2013, The shift, the growth and the promise. Available at: http://www.ey.com/Publication/vwLUAssets/Turkey_attractiveness_survey_2013/$FILE/turkey_at tractiveness_2013.pdf.

European Commission/High representative of the European Union for Foreign Affairs and Security Policy, 2012. Joint Staff Working Document – SWD(2012) 110 final – Implementation of the European Neighbourhood Policy in Armenia, Progress in 2011 and recommendations for action. Brussels, 15 May 2012. Available at: http://www.enpi-info.eu/library/content/armenia-enp-progress-report-2011.

European Commission (EC), 2011. Communication from the Commission to the European Parliament and the Council, Renewable Energy: Progressing towards the 2020 target, SEC(2011) 129 final, SEC(2011) 130 final, SEC(2011) 131 final-COM(2011) 31 final, Brussels, 31 January 2011. Available at: http://eur-lex.europa.eu/LexUriServ/LexUriServ.do?uri=COM:2011:0031:FIN:EN:PDF.

European Commission, 2009. Commission Staff Working Document, 2009 Annual Report of the market Observatory for Energy. Brussels, 21 December 2009. SEC(2009) 1734 final. Available at: http://www.uni-mannheim.de/edz/pdf/sek/2009/sek-2009-1734-en.pdf.

Eurostat, 2012. Statistics in focus, 44/2012. Renewable energy – Analysis of the latest data on energy from renewable sources. Available at: http://epp.eurostat.ec.europa.eu/cache/ITY_OFFPUB/KS-SF-12-044/EN/KS-SF-12-044-EN.PDF (It concerns the EU Member States).

Eurostat, 2013 http://epp.eurostat.ec.europa.eu/portal/page/portal/energy/introduction.

Ganev Peter, 2009. *'Bulgarian electricity market restructuring'*. Utilities Policy 17: 65–75.

Frankfurt School – UNEP Collaborating Centre for Climate & Sustainable Energy Finance, 2012. Global Trends in Renewable Energy Investment 2012. At: http://fs-unep-centre.org/sites/default/files/publications/globaltrendsreport2012final.pdf.

German Federal Environment Agency (Umweltbundesamt), 2009. Role and Potential of Renewable Energy and Energy Efficiency for Global Energy Supply. ISSN 1862-4359. At: http://www.ecofys.com/files/files/report_role_potential_renewable_energy_efficiency_global_energy_supply.pdf.

Greenpeace, 2012. Energy [r]evolution – A sustainable world energy outlook. Partners: Greenpeace International, European Renewable Energy Council (EREC), Global Wind Energy Council.

(GWEC). Date: July 2012. ISBN: 978-90-73361-92-8. Available at: http://www.greenpeace.org/international/Global/international/publications/climate/2012/Energy%20Revolution%202012/ER2012.pdf.

IEA, 2013. http://www.iea.org/publications/freepublications/publication/RedrawingEnergyClimateMap_2506.pdf.

IEA, 2012. World Energy Outlook 2012 – Renewable Energy Outlook. Available at: http://www.worldenergyoutlook.org/media/weowebsite/2012/WEO2012_Renewables.pdf.

IFC, 2012. Analysis of Feed-in Tariff for Renewable Energy Sources in Armenia. Available at: http://r2e2.am/wp-content/uploads/2012/08/FiT-analysis_English.pdf.

IFC, 2011b. Renewable Energy Policy in Russia: Waking the Green Giant. Available at: http://www1.ifc.org/wps/wcm/connect/region__ext_content/regions/europe+middle+east+and+north+africa/ifc+in+europe+and+central+asia/publications/renewable+energy+policy+in+russia+-+waking+the+green+giant or http://www1.ifc.org/wps/wcm/connect/bf9fff0049718eba8bcaaf849537832d/PublicationRussiaRREP-CreenGiant-2011-11.pdf?MOD=AJPERES.

IFC/GEF, 2012. Financing Renewable Energy Investments in Russia: Legal Challenges and opportunities. IFC Advisory Services in Europe and Central Asia, Russia Renewable Energy Program. Available at: http://www.ifc.org/wps/wcm/connect/9f6e55804df642c69c35bc7a9dd66321/PublicationRussiaRREP-Financing Energy.pdf?MOD = AJPERES.

IMF, 2012. World Economic Outlook – Growth Resuming, Dangers Remain – April 2012. World Economic and Financial Surveys. Available at: http://www.imf.org/external/pubs/ft/weo/2012/01/pdf/text.pdf.

IPA Energy + Water Economics, 2010. Study on the Implementation of the New EU Renewables Directive in the Energy Community- Final Report to Energy Community Secretariat. June 2010. At: http://www.energy-community.org/pls/portal/docs/644177.PDF.

InvestEast, 2010. *An overview of the Renewable Energy market in Romania*. Available at: http://investeast.ro/renewable_energy_in_romania.pdf.

Jefferson Institute, 2009. Serbia's Capacity for Renewables and Energy Efficiency. Available at: http://www.jeffersoninst.org/sites/default/files/serbia_policy_paper_0_0.pdf.

Konidari P. & Mavrakis D., 2006. Multi-criteria evaluation of climate policy interactions. Journal of Multi-Criteria Decision Analysis 14, pp. 35–53.

Konidari P. & Mavrakis D., 2007. A multi-criteria evaluation method for climate change mitigation policy instruments. Energy Policy 35, pp. 6235–6257.

KPMG, 2010. Energy & Natural Resources – Taxes and Incentives for Renewable Energy. At: http://www.enerclub.es/files/frontAction.do;jsessionid = 5863EC4D272BEDB765B5B220CF658CC0?action = getFile&fileID = 1000073675.

Global Wind Energy Council (GWEC), 2011. *Global Wind Report – Annual market update 2010*. Available at: www.gwec.net.

Lefkowitz Kenneth, 2009. *Mobilizing Investment in Renewable Energy in Bulgaria*. Presented at 5th International Congress for South-East Europe – Energy Efficiency & Renewable Energy Sources, 6–8 April 2009. Available at: http://necadvisory.com/uploads/Mobilizing%20Investment%20in%20Renewable%20Energy%20in%20Bulgaria.pdf.

Mattoo Aaaditya, Subramanian Arvind, 2011. World Development Vol. 40, Issue 6, pp. 1083–1097.

Mihailescu Loredana, 2009. *Wind Power in Romania*. Renewable energy focus, July/August 2009.

MOIEEP, 2011. Ministry of Economic Affairs and Communications, 2011. Mid-term overview of implementation of Energy Efficiency Plan 2007–2013 and further implementation. The second energy efficiency action plan of Estonia. Available at: http://ec.europa.eu/energy/demand/legislation/doc/neeap/estonia_en.pdf.

Nitoiu Maria, 2011. *Romania: Renewable Energy market Profile*, U.S. Commercial Service, United States of America, Department of Commerce, available at: http://www.nevadadec.com/Expotech/ElectraTherm/Romania%20-%20%20Renewable%20Energy%20Market%20Profile.pdf.

OECD, 2012. Attracting Investment in Renewable Energy in Ukraine. Available at: http://www.oecd.org/countries/ukraine/UkraineRenewableEnergy.pdf.

Patlitzianas Konstantinos, Karagounis Konstantinos, 2011. *The progress of RES environment in the most recent members states of the EU*. Renewable Energy 3: 429–436.
Pilifilip, 2012. EU Renewable Regime & Incentive Guidebook – Romania, 1 April 2012. At: http://www.pelifilip.com/wp-uploads/publications/cb1a6-publication.pdf.
PWC, 2012. Turkey's Renewable Energy Sector from a Global Perspective. At: http://www.pwc.com.tr/tr_TR/tr/publications/industrial/energy/assets/Renewable-report-11-April-2012.pdf.
Republic of Moldova, 2012. National Report for United Nations Conference on Sustainable Development 2012 Rio + 20. At: http://www.uncsd2012.org/content/documents/782Moldova_Report_RIO20_ENG_12-06-2012_final.pdf.
Republic of Serbia, 2013. National Renewable Energy Action Plan of the Republic of Serbia in accordance with the template as per Directive 2008/29/EC (Decision 2009/548/EC). Available at: http://www.ekapija.com/dokumenti/nreap_010313.pdf.
Republic of Serbia, 2010. Biomass Action Plan for The Republic of Serbia 2010 – 2012. Available at: http://www.woodybiomass.org/PagesRS/www.woodybiomass.org/userfiles/files/BAPenglish.pdf.
Romanian Business Digest, 2013. Wind Energy and other renewable energy sources in Romania. At: http://mcr.doingbusiness.ro/uploads/51cd884937ad7TPA%20Horwath_Wind%20Energy%20and%20other%20renewable%20energy%20sources%20in%20Romania.pdf.
Ruslan Stefanov, Denitza Mantcheva, Nikolay Tagarov, Dr. Dobromir Hristov, & Valentina Nikolova, 2011. *'Green Energy Governance in Bulgaria at a crossroads'*. Center for the Study of Democracy, ISBN: 987-954-477-174-4.
Pirlogea Corina, 2011. *'Barriers to investment in Energy from Renewable Sources'*. Economia, Seria management, Vol. 14, Issue 1. Available at: http://www.management.ase.ro/reveconomia/2011-1/12.pdf (last visited: 7 October 2011).
SEI, 2012a. Energy for a Shared Development Agenda: Global Scenarios and Governance Implications. Authors: Måns Nilsson, Charles Heaps, Åsa Persson, Marcus Carson, Shonali Pachauri, Marcel Kok, Marie Olsson, Ibrahim Rehman, Roberto Schaeffer, Davida Wood, Detlef van Vuuren, Keywan Riahi, Branca Americano and Yacob Mulugetta. Available at: http://www.sei-international.org/mediamanager/documents/Publications/SEI-ResearchReport-EnergyForASharedDevelopmentAgenda-2012.pdf.
SEI, 2012b. http://www.sei-us.org/Publications_PDF/SEI-LEAP-brochure-Jan2012.pdf.
Sirin S.M., Ege A., 2012. Overcoming problems in Turkey's renewable energy policy: How can EU contribute?, Renewable and Sustainable Energy Reviews, Vol. 16, Issue 7, September 2012, pp. 4917–4926.
Tesic Milos, Kiss Ferenc & Zavargo Zoltan, 2011. Renewable energy policy in the Republic of Serbia. Renewable and Sustainable Energy Reviews 15, pp. 752–758.
Treaty of Accession. 2003. Negotiations on Accession by the Czech Republic, Estonia, Cyprus, Latvia, Lithuania, Hungary, Malta, Poland, Slovenia and Slovakia to the

European Union. Legislative Acts and Other Instruments. AA 2003 Final. Brussels, 3 April 2003. http://www.eipa.eu/files/repository/product/20070816125607_2003w03.pdf.

Trypolska Galyna, 2012. Feed-in tariff in Ukraine: The only driver of renewables' industry growth? Energy Policy 45, pp. 645–653.

Willems, S. & Anakhasyan, E., 2011. *Analysis for European Neighbourhood Policy (ENP) Countries and the Russian Federation of social and economic benefits of enhanced environmental protection – Armenia Country Report.* Available at: http://www.environment-benefits.eu/pdfs/Armenia-ENPI%20Benefit%20Assessment.pdf.

UNDP, 2010. Use of Renewable Energy Sources in the World and Armenia – Through innovations to clear Technologies. Available at: http://www.nature-ic.am/res/pdfs/projects/CP/SNC/RE%20Brochure_eng.pdf.

UNECE, 2010. Financing Energy Efficiency Investments for Climate Change Mitigation – Regional Analysis of Policy Reforms to promote energy efficiency and Renewable energy investments. At: http://www.unece.org/fileadmin/DAM/energy/se/pdfs/eneff/eneff_pub/EE21_FEEI_RegAnl_Final_Report.pdf.

UNECE, 2011. Summary on reports from the national experts on development of renewable energy in the Russian Federation and CIS countries. At: http://www.unece.org/fileadmin/DAM/energy/se/pdfs/eneff/RES_RF_CIS/SummaryNationalReports.pdf.

UNECE, Energy Efficiency 21 Programme, 2009b. Republic of Moldova: National Energy Policy Information for Regional Analysis. Available at: http://www.clima.md/public/102/en/EnergyPolicyInformationForRegionalAnalysisMoldova.pdf.

UNEP, 2011. Renewable Energy, Investing in energy and resource efficiency. At: http://www.unep.org/greeneconomy/Portals/88/documents/ger/GER_6_RenewableEnergy.pdf.

UNEP Risoe, 2012. CDM Pipeline. Available at: http://cdmpipeline.org/.

UNFCCC, 2011. Report of the in-depth review of the fifth national communication of Estonia. Note by the secretariat. CC/ERT/2011/13, 24 August 2011. Available at: http://unfccc.int/national_reports/annex_i_natcom/idr_reports/items/4056.php and http://unfccc.int/files/kyoto_protocol/compliance/plenary/application/pdf/cc-ert-2011-13_idr_of_nc5_of_estonia.pdf.

United Nations, Department of Economic and Social Affairs, 2011. World Population Prospects – The 2010 Revision, Volume I: Comprehensive Tables. Available at: http://esa.un.org/unpd/wpp/Documentation/pdf/WPP2010_Volume-I_Comprehensive-Tables.pdf and http://esa.un.org/unpd/wpp/Documentation/pdf/WPP2010_Volume-II_Demographic-Profiles.pdf.

USAID, 2010. Wind Energy Development in Armenia: Legal, Regulatory, Tax and Customs Regulations – Assistance to energy sector to strengthen energy security and regional integration. Contract number EPP-I-08-03-00008-00. Available at: http://www.armesri.am/Public_Docs/DR/Task2/Wind%20Development_Legal_Eng.pdf.

USAID, 2011. Overview on Solar Electric Power in Buildings with applications, Armenia – Assistance to energy sector to strengthen energy security and regional integration. Contract Number EPP-I-08-03-00008-00. This publication was produced for review by the United States Agency for International Development. It was prepared by Tetra Tech ES, Inc. Available at: http://www.armesri.am/Public_Docs/DR/Task2/Overview%20on%20solar%20electric%20power%20in%20buildings.pdf.

Vaillancourt Kathleen, Waaub Jean-Philippe, 2004. Equity in international greenhouse gases abatement scenarios: A multicriteria approach. European Journal of Operational Research 153, pp. 489–505.

Valdur Lahtvee, 2012. Resource Efficiency Gains and Green Growth Perspectives in Estonia. http://library.fes.de/pdf-files/id-moe/09351.pdf.

World Economic Forum, 2013. Scenarios for the Russian Federation – World Scenario Series. Available at: http://www3.weforum.org/docs/WEF_Scenarios_Russian Federation_Report_2013.pdf.

CHAPTER 9
# Transformation of German- and European-Style Feed-In Tariff Schemes in East Asia in the Post-Fukushima Age: Recent Developments in Japan, South Korea, and Taiwan[*]

*Anton Ming-Zhi Gao & Chien Te Fan*

## §9.01 INTRODUCTION

Faced with an energy crisis and then climate change since the 1970s, the EC (and afterward the EU) has adopted a wide range of measures to promote the development of renewable electricity. Since then, many mechanisms have been implemented, including investment subsidies, tax incentives, feed-in tariffs (FITs), and net-metering. Finally, after the competition between FIT and renewable portfolio standard (RPS), FIT emerged as the prevailing way to facilitate the deployment of renewable electricity technology, as witnessed by the recent booming development in Germany. Long before the Fukushima accident, a group of advanced Asian countries had pondered both RPS and FIT and whether to introduce such a successful regime in their own countries, but instead, since the early 2000s, they adopted an 'adapted' version of the promotion scheme between the FIT scheme and RPS. Japan set up an RPS scheme (by Act on Special Measures Concerning New Energy Use by operators of electric utilities) in 2002, combined with a voluntary net-metering scheme, but afterward introduced an adapted form of FIT, that is, mandatory net-metering in photovoltaic (PV) in 2009. Then, in

---

[*] The authors would like to acknowledge the support funding from National Science Council of Taiwan. (Project number: NSC101-2410-H-007-024-MY2;102-3113-P-007-002) A little Part of this article is adapted from Taiwan Chapter of International Encyclopedia of Laws.

response to Fukushima, Japan adopted a combined regime of FIT and a net-metering scheme in 2011, which will come into effect in the middle of this year. South Korea, on the other hand, adopted a FIT scheme in 2001 and 2002 under the Electricity Utility Act and Promotional Act of NRE Development, Utilization and Deployment (2nd/3rd Amendment, but then decided to introduce an RPS scheme in 2010, which will also come into effect this year. In addition, Taiwan introduced a restricted version of the FIT scheme by a special guideline published by the state-owned power company in 2003, and a more aggressive version in 2009. Since 2011, however, Taiwan has adopted a combined regime of FIT and tendering scheme.

The purpose of this article, then, is to observe the *transitional* track of development of these legal regimes concerning the renewable 'electricity' sector in these three Asian countries and then compare the adapted form of the FIT schemes in these three countries with the mainstream German model. Most literature focuses on the development and evolvement of the renewable electricity legal regime 'within' each country, but there is a lack of comparative analysis, not to mention the creation of parameters to compare sub-mechanisms (such as the eligibility, eligible period, tariff schedule, tariff degression, tariff rise in response to the consumer price index, hard or soft cap, funding sources, and so on) under the FIT scheme with the European and German models. During the evolution of different types of regime, it would be interesting to see how the prevailing German FIT is introduced, transformed, and redefined in these Asian countries.

This article might be one of the few newer articles that provide an overview of the latest Asian renewable electricity regime after the Fukushima accident. Special attention will be paid to the special status of the PV regime in these three countries: Taiwan, a tendering; Japan, a mixed of net-metering and FIT; and South Korea, a preferred RPS scheme with mandatory installation capacity for PV. Therefore, the aim of this article is to contribute to the academic literature on the latest renewable electricity legal regime.

Possible reasons for this 'third way' of promoting renewable electricity in the aforementioned Asian countries are also examined. Unlike the ambitious European and lenient US approaches in promoting renewable electricity, for some reason Japan, South Korea, and Taiwan are still wavering between FIT and RPS, which could be due to economic reasons. Because they rely heavily on the manufacturing sector, facilitating the domestic PV application and boosting PV export creates a dilemma. Add to that the increasing costs of electricity. Therefore, the main focus remains on boosting PV export. Yet, without the domestic application, the government still worries about the image of climate change opportunists. It could be the transitional geological and political features and legal culture, as these countries are usually a mixture of Anglo-Saxon- and European-style economic and legal regimes, which leads to the conflict between a truly US-style RPS and a German-style FIT. Thus, the goal of this article is to further investigate the reasons for this particular approach in promoting renewable electricity.

## §9.02 DEVELOPMENT OF PRIMARY RENEWABLE ELECTRICITY PROMOTION SCHEMES IN JAPAN, SOUTH KOREA, AND TAIWAN

### [A] Japan

Based on the intensity of measures used to promote renewable electricity, Japan's promotion scheme can be divided into four stages.

#### [1] Phase I: Preparatory Stage: Prior to 1997

During this first stage, there was no legal instrument in place; thus, promotion of renewable electricity was strongly affected by the political will of the ruling parties. For instance, in 1994, Japan adopted the *'Basic Guideline for New Energy Introduction'* as a Cabinet decision, setting out the government's position on new and renewable energy for the first time. In May 1997, the Cabinet adopted an *'Action Plan for the Reform and Creation of Economic Structures'* to initiate structural reform of the Japanese economy.[1] In addition, to support technology development, Japan also provide a *Budget for Renewable Electricity RD&D Projects*. Yet, without having laws in place, particularly to promote the use of renewable electricity, these measures were unlikely to lead to substantial development.

#### [2] Phase II: 1997–2002

To further advance the development of renewable electricity, in 1997 a law – the *Law Concerning the Promotion of the Use of New Energies*[2] – was promulgated in June 1997. However, this law was defined as a 'framework.'[3] Its status as a framework is reflected in language referring to the *'planning,'* *'suggestive,'* and *'organizational'* approach under this law, rather than providing a more direct subsidy scheme.[4]

The *planning* element places an obligation on government and business operators to conduct some planning work. With respect to government planning duties, Article 3(1) of this law obliges the Minister of Economy, Trade, and Industry to prescribe and publicize the *Basic Policy* (hereinafter referred to as 'Basic Policy') for promoting new energy utilization; the New Energy Utilization Guidelines under

---

1. IEA, ENERGY POLICIES OF IEA COUNTRIES: JAPAN 2003 97 (2003).
2. Act on the Promotion of New Energy Usage (Act No. 37 of 18 Apr. 1997), available at: http://www.japaneselawtranslation.go.jp/law/detail_main?re=2&vm=&id=1892 (visited on 1 Sep. 2012).
3. IEA, *supra* note 1, at 99.
4. For example, RPS, Feed in tariff, net-metering, tax scheme. *See* e.g., ORGANISATION FOR ECONOMIC CO-OPERATION AND DEVELOPMENT & INTERNATIONAL ENERGY AGENCY, RENEWABLE ENERGY: MARKET & POLICY TRENDS IN IEA COUNTRIES 85 (2004).

Article 5 of this law also reflected this planning nature.[5] Similar planning duties are mandated in Chapter III for business operators.[6]

The *suggestive* nature of this law indicates its non-regulatory side. The aforementioned planning serves only to provide 'recommendations' for the actions of local government,[7] energy users,[8] business operators,[9] etc. If necessary, the competent authorities have only the power to give 'advice'[10] to energy users, rather than enforcement authority.

Finally, for the *organizational* element, a professional 'administrative legal person' – The New Energy and Industrial Technology Development Organization (NEDO)[11] – was entrusted with providing 'loan guarantees'[12] and other related duties.[13] Furthermore, small and medium size (SME) companies were encouraged to participate in renewable energy-related activities.[14]

---

5. Article 5:'(1) In order to promote New Energy Utilization, etc. by Energy Users for whom New Energy Utilization, etc. is deemed to be appropriate based on the circumstances, such as the characteristics of New Energy Utilization, etc., the technical level related to New Energy Utilization, etc., the Minister of Economy, Trade and Industry, taking these matters into consideration and giving due consideration to the preservation of the environment, shall prescribe and publicize New Energy Utilization, etc. guidelines for Energy Users (hereinafter referred to as *"New Energy Utilization Guidelines"*), relating to the types and methods of New Energy Utilization, etc. to be promoted.
   (2) The Minister of Economy, Trade and Industry shall revise the *New Energy Utilization Guidelines* when necessary due to a change in the circumstances set forth in the preceding paragraph.
   (3) The Minister of Economy, Trade and Industry shall consult in advance with the heads of relevant administrative organs, when he/she intends to prescribe or revise *the New Energy Utilization Guidelines*.'
6. Chapter III Promoting New Energy Utilization, etc. as Carried out by Business Operators, Art. 8: '(1) Business operators who intend to practice New Energy Utilization, etc. in their business activities (including those who intend to establish a juridical person in order to practice said New Energy Utilization, etc.) shall prepare a plan concerning said New Energy Utilization, etc. (hereinafter referred to as *"Utilization Plan"*) and submit said plan to the competent minister and may receive an accreditation to the effect that said Utilization Plan is suitable.'
7. Article 7: 'In establishing and enforcing policies that contribute to promoting New Energy Utilization, etc. in local areas, local government shall *take into consideration the Basic Policy* as much as possible.'
8. Article 4: '(1) Energy Users shall endeavor to achieve New Energy Utilization, etc., while giving due consideration to the provisions of the Basic Policy.'
9. Article 4: '(2) Energy Supply Business Operators and Manufacturing Business Operators, etc. *shall endeavor to* promote New Energy Utilization, etc., while giving due consideration to the provisions of the Basic Policy.'
10. Article 6: 'When the competent minister finds it to be necessary to promote New Energy Utilization, etc., he/she shall provide guidance and *advice* to Energy Users with regard to the matters provided for in the New Energy Utilization Guidelines.'
11. The introduction of NEDO, please refer to this website: NEDO, About NEDO, available at: http://www.nedo.go.jp/english/introducing_index.html (visited on 1 Sep. 2012).
12. Article 10: '(i)Guarantee debts pertaining to the funds required for New Energy Utilization, etc. which is done by a Certified Business Operator in accordance with a Certified Utilization Plan.'
13. Article 10: '(ii)Carry out businesses incidental to the business listed in the preceding item.'
14. Article 13 (Special Provisions of the Small and Medium Business Investment & Consultation Companies Act).

Chapter 9: Transformation of FIT in Japan, South Korea, and Taiwan    §9.02[A]

*[3]    Phase III: RPS Law and Voluntary Net-Metering Scheme*

*[a]    RPS Law in 2003*

Due to the relatively ineffective nature of measures to promote renewable electricity under the 1997 law, a more powerful and more direct scheme to promote renewable electricity (the *Special Law on the Use of New Energy by Electric Utilities*)[15] was promulgated in 2002 and came into force in 2003.[16] Under this law, an RPS scheme was set up to advance the previously weak renewable promotion scheme. Under this RPS clause, certain 'electricity utilities' (including electric power companies, specified electric utilities, and specified-scale electric utilities, but excluding industrial self-generators) are obliged to use a minimum amount of new energy sources[17] and must report on the results of their efforts to meet this requirement during the previous fiscal year by 1 June of each year.[18] The percentage of new energy sources required will be redefined annually. According to the portfolio standard, retailers were obliged to acquire approximately 1.35% of their sales volume from new energy sources in 2010.[19] There are three ways to fulfill this requirement: (1) generate their own green electricity, (2) buy green electricity from other companies that generate electricity from renewables, and/or (3) buy a portion of the obligations fulfilled by other retailers through purchase of an 'applicable amount of New Energy Electricity.'[20]

The definition of renewable electricity requires further explanation. There is no minimum percentage requirement for specific types of renewable energy to meet the standard.[21] At first glance, this law appears to include most types of renewable

---

15. Act on Special Measures Concerning New Energy Use by operators of electric utilities (The revision of Art. 2, paragraph (2), item (vi) has not come into force.) Act No. 62 of 7 Jun. 2002 (電気事業者による新エネルギー等の利用に関する特別措置法 (第二条第二項第六号改正未施行) 平成十四年六月七日法律第六十二号), available at: http://www.japaneselawtranslation.go.jp/law/detail/?id=1889&vm=04&re=02 (visited on 1 Sep. 2012).
16. IEA, ENERGY POLICIES OF IEA COUNTRIES: JAPAN 2008, 152 (2008).
17. Article 4: '(1)Operators of electric utilities shall, pursuant to Ordinance of the Ministry of Economy, Trade and Industry, notify the Minister of Economy, Trade and Industry by 1 June of each year of their *Standard Amount of Use of Electricity from New Energy*, etc. (meaning the amount of Electricity from New Energy, etc. calculated pursuant to provisions of Ordinance of the Ministry of Economy, Trade and Industry as the amount to be used by said operators of electric utilities in the relevant notification year, in consideration of the prevalence of generation facilities for voltage regulation that become necessary as a result of having set a *Use Target for Electricity from New Energy, etc.* and installed Generation Facilities for Electricity from New Energy, etc. based on the amount of electricity supplied by said operators of electric utilities in the fiscal year preceding the notification year (except for the electricity supplied to other operators of electric utilities the same shall apply in Art. 10); the same shall apply hereinafter), which they plan to use for the one year period from 1 April of the relevant year to 31 March of the following year (hereinafter referred to as the "Notification Year"), and other matters as prescribed by Ordinance of the Ministry of Economy, Trade and Industry.'
18. Article 10: 'Pursuant to the Ordinance of the Ministry of Economy, Trade and Industry, operators of electric utilities *shall notify* the Minister of Economy, Trade and Industry by 1 June of each year of the amount of electricity supplied from 1 April of the preceding year to 31 March of the current year, and of other matters specified by Ordinance of the Ministry of Economy, Trade and Industry.'
19. IEA, *supra* note 1, at 99.
20. *Ibid.*, at 99.
21. IEA, *supra* note 16, at 152.

electricity as eligible to meet these obligations.[22] Yet, in fact, there are only *limited types of renewable electricity* admitted that qualify.[23] In general, wind power, biomass power, and cogeneration qualify as renewable electricity, but certain limitations or criteria are provided for geothermal, hydroelectric, and PV. Large hydroelectric (≥1,000 kW), conventional flash-type geothermal, and geothermal energy affecting the hot water supply are excluded from this scheme.[24] In addition, PV systems awarded subsidies,[25] i.e., 'designated photovoltaic electricity,' are also excluded; otherwise, their contribution as a 'renewable' element would be double-counted.

To manage this scheme, a renewable certificate program was also established under Article 9 of the Act.[26] In FY 2009, the number of power generation facilities utilizing new energy sources totalled 83,562, with installed capacity equivalent to

---

22. Article 2: '(2) The term *"New Energy, etc."* as used in this Act means the following types of energy:
    (i) wind power; (ii) photovoltaic power; (iii) geothermal power; (iv) hydroelectric power *(limited to that as specified by Cabinet Order)*; (v) heat produced with biomass (organic substances derived from animals and plants which can be used as energy sources (excluding crude oil, petroleum gas, combustible natural gas, coal, and products manufactured therefrom)); (vi) in addition to the types of energy listed in the preceding items, energy other than heat produced with fossil fuels (including crude oil, petroleum gas, combustible natural gas, coal, and products manufactured therefrom (including the by-products of such manufacture supplied for incineration)) as *specified by Cabinet Order.*
    (3) The term *"Electricity from New Energy, etc."* as used in this Act means electricity that is generated through the conversion of New Energy, etc. using Facilities for Generating Electricity from New Energy, etc.'
23. Article 2(4): 'The term "Facilities for Generating Electricity from New Energy" as used in this Act means facilities that convert new energy, etc. into electricity and that have been certified pursuant to the provisions of paragraph (1) of Article 9.'
24. S. Kawazoe, Geothermal Japan - Geothermal Resources Council, available at: geothermal.org/articles/GeoJapan.pdf (visited on 1 Sep. 2012).
25. 'Designated photovoltaic electricity' refers to electricity which the electric utilities are obliged to purchase under the 'New Purchase System for Photovoltaic Electricity,' which comprises part of the electricity from new energy sources generated at photovoltaic power generating facilities. This electricity may not be applied to fulfillment of obligation under the RPS Act. The amount of designated photovoltaic electricity supplied in FY 2009 is for the period after implementation of the New Purchase System for Photovoltaic Electricity (i.e., from 1 Nov. 2009 to 31 Mar. 2010). The amount before implementation of the system (from 1 Apr. 2009 to 31 Oct. 2009) is included in the amount supplied via photovoltaic power generation.
26. Article 9: '(1) A person who generates or intends to generate electricity using a facility that converts New Energy, etc. into electricity may receive certification from the Minister of Economy, Trade and Industry pursuant to the provisions of Ordinance of the Ministry of Economy, Trade and Industry through conformity with both of the following items:

    (i) the facility for converting New Energy, etc. into electricity that is installed or planned for installation by the person who generates or intends to generate said electricity, conforms to the standards prescribed by Ordinance of the Ministry of Economy, Trade and Industry; and
    (ii) the method of generating electricity conforms to the standards prescribed by the Ordinance of Ministry of Economy, Trade and Industry.

    (2) When the generation of electricity pertaining to an application for the certification in the preceding paragraph is found to conform to all of the items in the preceding paragraph, the Minister of Economy, Trade and Industry shall issue certification pursuant to said paragraph.
        (3) When the Minister of Economy, Trade and Industry intends to issue certification pursuant to paragraph (1) concerning a Generation Facility for Electricity from New

645,238 kW; these facilities were certified pursuant to Article 9 of the RPS Act. As of the end of FY 2009, a total of 519,966 power generation facilities utilizing new energy sources had been certified, and cumulative installed capacity had reached 6,486,158 kW (compared to 5,853,731 kW as of the end of FY 2008).[27]

An implementation report for RPS is required of the obligatory electricity utilities and is published on the METI website. For example, information for 2010 is provided below (Table 9.1).[28]

In FY 2010, the RPS Act required 53 electric utilities (10 electric power companies, 5 specified electric utilities, and 38 specified-scale electric utilities) to utilize a total of 11,014,697,000 kWh of renewable-energy-sourced electricity.

The total amount of renewable-energy-sourced electricity that was supplied to electric utilities from renewable energy power generation facilities was 10,245,907,844 kWh in FY 2010 (8,873,162,050 kWh in FY 2009). This includes 1,336,893,027 kWh of designated PV electricity that does not count toward fulfilment of the requirements.

Table 9.1  Total Amount of Renewable Electricity in 2010 under the RPS Scheme

| Total Amount of Renewable-Energy-Sourced Electricity (in kWh) | |
|---|---|
| Power Generation Source | Electricity Supplied |
| Wind power | 4,143,413,522 |
| Hydroelectricity | 989,079,899 |
| Photovoltaic power | 16,615,435 |
| Biomass power | 3,744,516,697 |
| Geothermal power | 10,535,868 |
| Cogeneration | 4,853,396 |
| Total | 8,909,014,817 |
| Designated photovoltaic power* | 1,336,893,027 |
| Grand Total | 10,245,907,844 |

Source: METI, Report on the Implementation of the RPS Act in FY 2010, available at http://www.meti.go.jp/english/press/2011/0714_01.html. (visited on 1 September 2012).

(Reference) Banked power carried over from FY 2009 to FY 2010: 6,405,731,000 kWh.

---

Energy, etc., he/she shall consult with the Minister of Agriculture, Forestry and Fisheries, the Minister of Land, Infrastructure and Transport, or the Minister of Environment in advance, pursuant to the provisions of Cabinet Order,
    (4) The Minister of Economy, Trade and Industry may revoke certification pursuant to paragraph (1) when the generation of electricity pertaining to said certification is found to no longer conform to any individual item under said paragraph.
    (5) In addition to the matters prescribed in each of the preceding paragraphs, other matters necessary for certification pursuant to paragraph (1) shall be prescribed by Cabinet Order.'
27. METI, FY 2009 Implementation Status of the Act on Special Measures Concerning New Energy Use by Electric Utilities, available at: http://www.meti.go.jp/english/press/data/20100715_03.html (visited on 1 Sep. 2012).
28. METI, Report on the implementation of the RPS Act in FY 2010, available at: http://www.meti.go.jp/english/press/2011/0714_01.html (visited on 1 Sep. 2012).

* 'Designated photovoltaic power' refers to renewable-energy-sourced electricity produced at photovoltaic power generating facilities that the electric utilities are obliged to purchase under the 'Excess Electricity Purchasing Scheme for Photovoltaic Power.' This electricity may not be applied to fulfillment of obligations under the RPS Act.

*[b]    Voluntary Net-Metering Scheme*

Under the RPS law, each electric utility is required to purchase certain renewable electricity. To meet this requirement, electric utilities developed some 'voluntary' programs to fulfill their 'mandatory' obligations,[29] purchasing surplus electricity produced by *PV and wind energy* installers. The purchase price is set in accordance with the market price for electricity. Thus, the purchase rate schedule for electricity from residential customers is around JPY 21–24, and JPY 11–12 from enterprise or industrial customers. In 2006, the rate paid for solar PV was JPY 19–23/kWh.[30] However, it should be noted that only electricity purchased by the electric utilities can be counted as part of the RPS obligation, not that consumed by the customers themselves.

Due to the low purchase price, the outcome of this scheme was foreseeable. This approach is best for wind power, but not as good for high-cost PV. To facilitate PV development, additional measures were needed.

*[c]    Other Incentives*

[i]    Government Budget, Projects, and Programs

Japanese has used government funding to encourage development of renewable energy technology by injecting money into certain projects and programs. The budget for promotion of new energy (JPY 144.9 billion in FY 2002) has more than doubled over the past five years, and is JPY 34.3 billion greater than in the previous fiscal year, targeted toward existing programs and measures to reduce costs through technology improvements.[31]

[ii]    Taxes

The Japanese government uses the tax code to encourage development of renewable electricity projects. Various tax breaks and incentives are provided for residential and non-residential sectors (Table 9.2).

---

29. 'some solar PV also receives direct assistance, similar to a feed-in tariff. Under the excess power purchasing menu, electric utilities voluntarily purchase excess power primarily from residential generators that self-supply and sell their excess power back to the grid.' See IEA, *supra* note 16, at 154.
30. IEA, *supra* note 16, at 154.
31. IEA, *supra* note 1, at 100.

*Table 9.2 Tax Measures in Japan*

| | Tax Measures |
|---|---|
| Residential | Mortgage tax breaks for new homes |
| | Tax breaks for remodelling homes to conserve energy |
| | Lower local property taxes for new energy projects |
| Non-residential | 7% Tax deduction (SMEs) |
| | Tax deduction or special depreciation allowance for acquisition of new energy facilities |
| | Immediate amortization |
| | Exception to the fixed asset tax |

Source: METI, *New Buyback Program for Photovoltaic Generation in Japan* (October 2009), available at http://www.meti.go.jp/english/policy/energy_environment/renewable/ref2001.html.

[iii]   Investment Subsidies and Demonstration Project Subsidy

To reduce costs for renewable electricity developers, both central and local governments provide investment subsidies for certain types of renewables, mainly PV and wind power (Table 9.3). Funding is also frequently available for biomass[32] and small-scale hydropower facilities.[33]

*Central government.* At the central government level, in FY 2002, subsidies paid to PV plants for residential use (output < 10 kW) were JPY 0.1 million/kW, equivalent to 14% of the installation cost.[34] For larger PV plants with output < 50 kW, subsidies were 33% of the installation costs for enterprises[35] and 50% for municipal entities. Subsidies for electricity generation from waste were 10% of the installation cost for plants with < 15% efficiency, and for more efficient plants, 33% of the installation cost for enterprises and 50% for municipal entities. Subsidies paid to renewables plants such as wind[36] and biomass were, in principle, 33% of the installation cost for enterprises and 50% for municipal entities.[37]

---

32. For example, Biomass Power Generation(Itoigawa-shi) Support for business operators. See METI, Japan's Promotion Measures for Introduction, available at:http://www.meti.go.jp/english/policy/energy_environment/renewable/ref1004.html (visited on 1 Sep. 2012).
33. For example, Small-scale hydropower generation (Tsuru-shi) Subsidy for developing small and medium power plants *see* METI, Japan's Promotion Measures for Introduction, available at: http://www.meti.go.jp/english/policy/energy_environment/renewable/ref1004.html (visited on 1 Sep. 2012).
34. For example, Housing PV Support project for housing PV introduction *see* METI, Japan's Promotion Measures for Introduction, available at:http://www.meti.go.jp/english/policy/energy_environment/renewable/ref1004.html (visited on 1 Sep. 2012).
35. For example, PV Power Generation (Kawasaki-shi) Support for business operators. See METI, Japan's Promotion Measures for Introduction, available at:http://www.meti.go.jp/english/policy/energy_environment/renewable/ref1004.html (visited on 1 Sep. 2012).
36. For example, Wind Power (Hokuei-cho) Regional new energy project. See METI, Japan's Promotion Measures for Introduction, available at:http://www.meti.go.jp/english/policy/energy_environment/renewable/ref1004.html (visited on 1 Sep. 2012).
37. IEA, *supra* note 1, at 102.

Table 9.3  *Investment Subsidies in Japan*

|  | Investment Subsidy |
|---|---|
| Residential | JPY 70,000/kW for systems priced at JPY ≤700,000/kW that meet quality guarantees and other requirements |
| Non-residential | Local governments, etc., 1/2 of the initial cost<br>Private business operators, 1/3 of the initial cost |

*Source*: METI, *New Buyback Program for Photovoltaic Generation in Japan* (October 2009), available at http://www.meti.go.jp/english/policy/energy_environment/renewable/ref2001.html.

*Local governments.* In addition to subsidies provided by the central government, prefectures, cities, towns, and villages have implemented their own subsidy systems to complement national subsidies.[38]

*Demonstration subsidies.* Finally, the government has also funded demonstration projects on a project by project basis, e.g., the large-scale demonstration test of Mega Solar (Hokuto-shi).[39]

### [4]  Phase IV: 2009 Mandatory PV Net-Metering

*[a]  PV Net-Metering Scheme*

The above scheme remains subject to the mercy of the electric utility company buying surplus electricity. In addition, the purchase price is not sufficiently favorable for PV. Thus, the incentive program was further revised in 2009. This scheme was introduced using the administrative power of the government to change the existing voluntary scheme to a mandatory PV one.[40] Under this approach, the investment security of PV installers is much enhanced. First, a 10-year-long surplus electricity purchase requirement for electric utilities was established.[41] Second, a substantially higher fixed purchase tariff was specified, nearly double that previously paid. The rate schedule is illustrated in Table 9.4.

---

38. IEA, *supra* note 1, at 102.
39. METI, Japan's Promotion Measures for Introduction, available at: http://www.meti.go.jp/english/policy/energy_environment/renewable/ref1004.html (visited on 1 Sep. 2012).
40. 'The Ministry of Economy, Trade, and Industry (METI) today issued directives relevant to the Law on the Promotion of the Use of Nonfossil Energy Sources and Effective Use of Fossil Energy Materials by Energy Suppliers (Law No. 72 of 2009), including ministerial ordinances and notifications.' *See* METI, *The New Purchase System for Photovoltaic Electricity will be Launched November 1, 2009*, available at http://warp.ndl.go.jp/info:ndljp/pid/1364125/www.meti.go.jp/english/press/data/20090831_02.html (visited on 1 Sep. 2012).
41. METI, *About the New Buyback Program for Photovoltaic Generation*, available at http://www.meti.go.jp/english/policy/energy_environment/renewable/ref2002_01.html. (visited on 1 Sep. 2012).

Table 9.4  Rate Schedule for PV in 2009–2012

|  | 11/2009–3/2011 (JPY/kW·h) | 2011–2012 (JPY/kW·h) |
|---|---|---|
| Household | Small PV ( < 10 kW): 48 (39)* | 42 (34)* |
|  | Large PV (≥10 kW): 24 (20)* | 40** (32)* |
| Other Non-household | 24 (20)* | 24*** (20)**** |

* Where a private electricity generator is also installed.

** For non-residential PV facilities in FY 2011. For customers who sign up in FY 2011, JPY 40 per kWh will be applied when confirmed by an RPS certificate and (1) subsidy for Acceleration to Install New and Renewable Energies has not been received and (2) PV facilities will be newly installed in FY 2011.

*** Facilities installed before FY 2010.

**** For double power generation.

*Source*: Compiled from http://www.meti.go.jp/english/policy/energy_environment/renewable/ref2002_01.html and http://www.meti.go.jp/english/press/2012/0301_02.html. (visited on 1 September 2012).

[b]   Other Incentives

However, to promote development of PV, simple reliance on the net-metering scheme remains insufficient; thus, a combination of net-metering with other programs,

Figure 9.1   Cost Recovery for Photovoltaic Generation Systems (for a New System: Model Case)

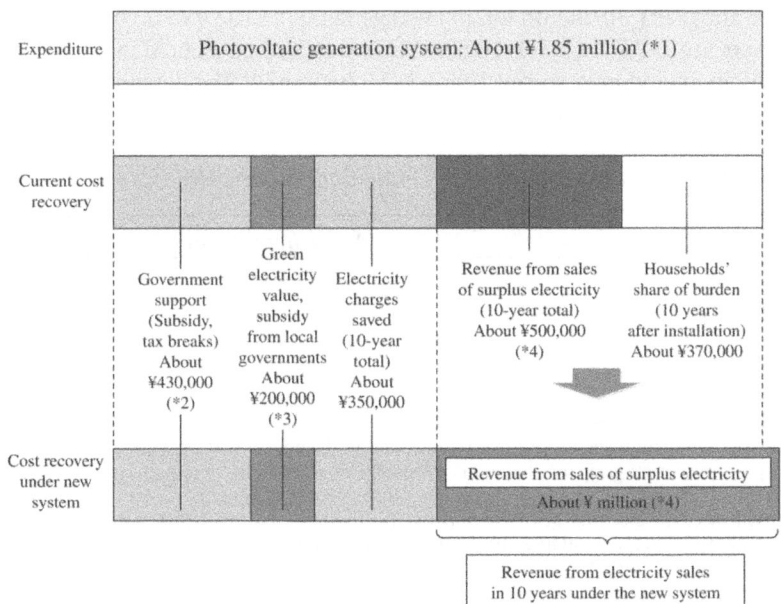

including investment subsidies from central and local governments and tax breaks, remains important. Figure 9.1 illustrates how the various programs work together.

(*1) The price of a photovoltaic generation system is an estimate based on applications for subsidies filed from January to March, 2009. Interest rates involved in system installation, maintenance costs, and repair costs after installation have not been taken into account.

(*2) Subsidy: JPY 70,000/1 kWh plus mortgage tax break (about JPY 190,000)

(*3) Revenue from sales of green electricity value (captive consumption) is calculated on the assumption that the value is about JPY 5 per 1 kWh. Separate contracts with certificate issuers are necessary. The amount and availability of subsidies depend on each local government. On average, the amount of subsidies provided by local governments (prefectural or municipal levels) is about JPY 3.8/kWh (FY 2008). (Example) The Tokyo Metropolitan government started a JPY 100,000/kWh subsidy system in April 2009.

(*4) The estimate is based on the following assumptions. Generating capacity: 3.5 kW, electricity sales ratio: about 60%, power generation efficiency: about 12%, unit price of electricity sold: JPY 24/kWh at present and JPY 48/kWh under the new system.

Source: METI, *About the New Buyback Program for Photovoltaic Generation (Buyback Price/Period)*, available at http://www.meti.go.jp/english/policy/energy_environment/renewable/ref2002_02.html (visited on 1 September 2012)).

[c]   *Passing the Costs on to Consumers*

Under the new program, the government allows the electric utility buying this more expensive electricity to pass the cost on to its customers. Recovery and transfer of buyback costs began in April 2010 (recovery of costs for 2009). The cost burden increases in proportion to the amount of electricity consumed. Initially, it costs about JPY 0.1/kWh, or about JPY 30/month for an average family.[42] The PV surcharge rates for electricity users are 1-7 sen (JPY 1/100)/kWh (a standard household using ~300 kWh/mon electricity would be assessed JPY ~3-21 /month).[43] The detailed table on PV Surcharge Rates is illustrated in Table 9.5.

Table 9.5   PV Surcharge Rates for Individual Electric Utilities

| Electric Power Company | FY 2011 PV Surcharge Rates (JPY/kW h) |
|---|---|
| Hokkaido | 0.01 |
| Tohoku | 0.03 |
| Tokyo | 0.03 |

---

42. METI, About the New Buyback Program for Photovoltaic Generation (Cost recovery), available at: http://www.meti.go.jp/english/policy/energy_environment/renewable/ref2002_03.html (visited on 1 Sep. 2012).
43. METI, Approval of electricity charges followed by the fixing of FY 2011 Photovoltaic Power Promotion Surcharge Rates, available at: http://www.meti.go.jp/english/press/2011/0126_01.html (visited on 1 Sep. 2012).

| Electric Power Company | FY 2011 PV Surcharge Rates (JPY/kW h) |
|---|---|
| Chubu | 0.06 |
| Hokuriku | 0.01 |
| Kansai | 0.03 |
| Chugoku | 0.06 |
| Shikoku | 0.06 |
| Kyushu | 0.07 |
| Okinawa | 0.06 |

PV surcharge rates are calculated based on the total expected demand for electricity in the current fiscal year. Any fraction less than 1 sen is excluded from the rate for that fiscal year, and the resulting shortfall is recovered by adjusting the surcharge for the following fiscal year.[44]

*[5] Phase V: Post-Fukushima: 2011 FIT with Mandatory Small PV Net-Metering*

After the Fukushima accident and closure and termination operation of several nuclear power plants, Japan faced a large-scale shortage of electricity and resulting electricity crisis. Thus, renewable energy was considered a priority for development. A more favorable scheme than the then-current mandatory net-metering that had been studied alongside mandatory net-metering in 2009 underwent more serious discussion.[45] The Fukushima accident played a key role in this more favorable approach gaining ground in 2011. After revision of the Act on Special Measures Concerning the Procurement of Renewable Energy by Electric Utilities (hereafter the '2011 FIT Law'),[46] a FIT scheme was introduced in July 2012, with mandatory net-metering remaining applicable to new residential PV systems < 10 kWh. This scheme can be considered an expansion of the previous PV net-metering scheme, as discussed below.

---

44. METI, Approval of electricity charges followed by the fixing of FY 2012 Photovoltaic Power Promotion Surcharge Rates, available at: http://www.meti.go.jp/english/press/2012/0125_02.html (visited on 1 Sep. 2012).
45. *See* METI, Present Status and Promotion Measures for the introduction of Renewable Energy in Japan, available at: http://www.meti.go.jp/english/policy/energy_environment/renewable/index.html (visited on 1 Sep. 2012).: 'On the basis of the study results by the "Project Team on Japan's Feed-in Tariff Scheme", which was established in November 2009, the Ministry of Trade, Economy and Industry (METI) drew up the framework (basic idea) of the Japan's feed-in tariff scheme for renewable energy. This material was studied at the fifth meeting of the above project team held on 23 Jul. 2010 and corrected with taking into consideration the opinions of knowledgeable people.'
46. Act on Special Measures Concerning the Procurement of Renewable Energy by Electric Utilities (電気事業者による再生可能エネルギー電気の調達に関する特別措置法), (平成二十三年八月三十日法律第百八号), available at: www.jurists.co.jp/ja/topics/docs/newsletter_201112_cb.pdf (visited on 1 Sep. 2012).

## [a] Detailed Comparison with PV Net-Metering Scheme

First, the renewable electricity types included have been expanded. Formerly, only PV was eligible, but now all types of renewables are included.[47] However, not all renewable electricity installations under each renewable energy type may benefit from this new favored scheme. For instance, hydropower cannot exceed 30,000 kW, while biomass power generation systems not affecting existing businesses such as the pulp and paper industry are eligible.[48]

Second, for eligible facilities, all electricity can be fed into the tariff, rather than just surplus electricity.

Third, there are mandatory grid connection[49] and contracting requirements[50] for electric utilities, an expansion of the former net-metering scheme. Only under certain circumstances can electric utilities refuse to buy, enter into a contract, or connect. The Act sets forth exceptions to the requirement that operators of electric utilities purchase the full amount of renewable electricity generated by the Specified Suppliers. Operators are excused from their obligation to enter into power purchase agreements and make related interconnections if there is 'a likelihood of unjust harm to the benefits of operators of electric utilities,' 'a likelihood of the occurrence of damage to securing the smooth supply of electricity,' or 'a just reason as set forth in the [Implementing Regulations].'[51] Finally, an electric power company may also be exempt from the obligation to connect if:

- The specified supplier does not bear the cost necessary for the connection as stipulated in the Ordinances of the Ministry of Economy, Trade and Industry.
- Electricity supply by the electric power company may be interrupted.
- There exists any good reason as stipulated in the Ordinances of the Ministry of Economy, Trade and Industry.[52]

---

47. Article 2 of 2011 FIT Law.
48. METI, Feed in Tariff Scheme for Renewable Energy, October 2011, available at: http://www.meti.go.jp/english/policy/energy_environment/renewable/pdf/summary201209.pdf (visited on 1 Sep. 2012).
49. Article 5 of 2011 FIT Law.
50. Article 4 of 2011 FIT Law.
51. Michael C. Graffagna and Yoshinobu Mizutani, Outline of Japan's Feed-In Tariff Law for Renewable Electric Energy, at p. 2, available at: http://www.mofo.com/files/Uploads/Images/110913-Outline-of-Japans-Feed-In-Tariff-Law-for-Renewable-Electric-Energy.pdf (visited on 1 Sep. 2012). See also, New Feed-in-Tariff System for Renewable Energy, October 2011, at p. 4, available at: http://www.lexology.com/library/detail.aspx?g=c4ec04a4-3a79-46d6-9157-757 83b894f38 (visited on 1 Sep. 2012): 'An electric power company may be exempt from the obligation to enter into a purchase agreement "if the terms may unduly prejudice the interests of the relevant electric power company or there exists any good reason as stipulated in the Ordinances of the Ministry of Economy, Trade and Industry".'
52. New Feed-in-Tariff System for Renewable Energy, October 2011, at p. 4, available at: http://www.lexology.com/library/detail.aspx?g=c4ec04a4-3a79-46d6-9157-75783b894f38 (visited on 1 Sep. 2012).

Fourth, under the former scheme the rate was set by the Ministry, but under the current scheme there is a more professional commission in charge of setting rates.[53] The rates are also more favorable than the previous scheme. For example, > 10 kW PV systems enjoy a rate of JPY 42/kWh, compared to the previous rate of 40. Taking into account payment for all electricity (rather than surplus electricity), 20 years (rather than 10 years), and that technology costs have decreased between 2011 and 2012, the new rate is substantially improved. The rates for other types of renewable electricity are also higher than under the former voluntary net-metering scheme. The detailed rate schedule[54] in Table 9.6 will be valid until the end of March 2013. METI is responsible for readjusting the tariffs and rates for future periods.[55]

Table 9.6  Rate Schedule for Feed-In Tariffs in 2012

| Renewable Tariffs in Japan | | | | | |
|---|---|---|---|---|---|
| 18 June 2012 | | | | | |
| | | | 103.374 | 1.299 | 1.315 |
| | Years | JPY/kWh | EUR/kWh | CAD/kWh | USD/kWh |
| Wind | 20 | | | | |
| < 20 kW | | 57.75 | 0.559 | 0.725 | 0.735 |
| > 20 kW | | 23.10 | 0.223 | 0.290 | 0.294 |
| Geothermal | 15 | | | | |
| < 15 MW | | 42.00 | 0.406 | 0.528 | 0.534 |
| > 15 MW | | 27.30 | 0.264 | 0.343 | 0.347 |
| Hydro | 20 | | | | |
| < 200 kW | | 35.70 | 0.345 | 0.448 | 0.454 |
| > 200 kW < 1 MW | | 30.45 | 0.295 | 0.383 | 0.387 |
| Photovoltaics | | | | | |
| < 10 kW for surplus generation | 10 | 42.00 | 0.406 | 0.528 | 0.534 |
| > 10 kW | 20 | 42.00 | 0.406 | 0.528 | 0.534 |
| Biogas from sewage sludge and animals | 20 | 40.95 | 0.396 | 0.514 | 0.521 |
| Biomass (solid fuel incineration) | | | | | |
| Sewage sludge and municipal waste | 20 | 17.85 | 0.173 | 0.224 | 0.227 |
| Forest thinnings | 20 | 33.60 | 0.325 | 0.422 | 0.428 |

---

53. Articles 31–37 under Chapter V. Feed in tariff Setting Commission.
54. ERIC JOHNSTON, FEED-IN TARIFF New feed-in tariff system a rush to get renewables in play, Tuesday, 29 May 2012, available at:http://www.japantimes.co.jp/text/nn20120529i1.html (visited on 1 Sep. 2012).
55. Japan: Feed-in tariff scheme confirmed, 25 Jun. 2012, available at: http://www.sunwindenergy.com/news/japan-feed-tariff-scheme-confirmed. (visited on 1 Sep. 2012).

| Renewable Tariffs in Japan | | | | | |
|---|---|---|---|---|---|
| 18 June 2012 | | | | | |
| | | | 103.374 | 1.299 | 1.315 |
| | Years | JPY/kWh | EUR/kWh | CAD/kWh | USD/kWh |
| Whole timber | 20 | 25.20 | 0.244 | 0.317 | 0.321 |
| Construction waste | 20 | 13.65 | 0.132 | 0.171 | 0.174 |
| Approved 18 June 2012 | | | | | |
| Effective 1 July 2012 | | | | | |

*Source:* Paul Gipe, Japan Approves Feed in Tariffs, available at: http://www.renewableenergyworld.com/rea/news/article/2012/06/japan-approves-feed-in-tariffs (visited on 1 September 2012).

Finally, there is a more comprehensive cost recovery scheme in the new FIT Law than for the previous PV net-metering. The FIT Law deals with the cost recovery issue in Chapter III, addressing cost-sharing among different electric utilities (Articles 8–18), and Chapter IV, *The Authority of Cost Burden Adjustment (Procurement Price Calculation Committee)*,[56] to determine who should bear the surcharge. All electricity carries the surcharge by kW·h, other than (1) those affected by the Great East Japan Earthquake (exempted from surcharge payment from 1 July 2012 to 31 March 2013)[57] and (2) enterprises that consume a large amount of energy (exempted from payment of 80% or more of the surcharge).[58] An equal surcharge/kWh will be assessed nationwide. The government will determine the surcharge/kWh based on the results of the previous fiscal year. Because it is possible that the introduction timeframe for renewable energy may vary depending on the region, an organization to adjust the burden will be established. The surcharge collected by the electric utilities will first be collected by the cost-bearing adjustment organization (Procurement Price Calculation Committee), and then delivered to the electric utilities as grants proportional to the actual

---

56. Articles 8–18 of Chapter III Cost Sharing Among Electricity Utilities (電気事業者間の費用負担の調整); Art. 19-30 of Chapter IV *Authority of Cost Burden Adjustment* (費用負担調整機関).
57. Price Relief for Earthquake Victims – Electricity users of offices, residences and other facilities and equipment that were severely damaged in the Great East Japan Earthquake and who meet additional requirements (if any) provided for in the applicable government ordinances will not be invoiced the surcharge during the nine month period from 1 Jul. 2012 to 31 Mar. 2013.
    *See* Michael C. Graffagna and Yoshinobu Mizutani, Outline of Japan's Feed-In Tariff Law for Renewable Electric Energy, available at: http://www.mofo.com/files/Uploads/Images/110913-Outline-of-Japans-Feed-In-Tariff-Law-for-Renewable-Electric-Energy.pdf (visited on 1 Sep. 2012).
58. Price Relief for Industrial Users – A reduction in the surcharge of 80% or more is to be provided to business facilities whose annual electricity usage amount exceeds an amount to be set forth in the Implementing Regulations, upon application by a business operator whose ratio of electricity usage (in kWh) to sales volume (per 1,000 yen) (i) exceeds 8 times the average ratio in the manufacturing industry (if a manufacturer) or (ii) exceeds the average ratio in the non-manufacturing industry (if a non-manufacturer) by a factor to be determined in the Implementing Regulations see Michael C. Graffagna and Yoshinobu Mizutani, Outline of Japan's Feed-In Tariff Law for Renewable Electric Energy, available at: http://www.mofo.com/files/Uploads/Images/110913-Outline-of-Japans-Feed-In-Tariff-Law-for-Renewable-Electric-Energy.pdf (visited on 1 Sep. 2012).

purchase costs.⁵⁹ Therefore, compared with potentially varying regional renewable surcharges, under the new scheme, the surcharge will be the same for everyone.

*[b]    Results*

PV and wind power benefit greatly under this new scheme. By 2 July 2012, the government had approved 44 solar or wind power facilities with a combined output of 41,605 kW to join the system, according to an official at the Agency for Natural Resources and Energy, part of the industry ministry.⁶⁰ Under this scheme, the electricity price per kW would increase by JPY 0.22, which corresponds to an increase in the electricity bill for a ordinary family of JPY ~87/month, and an annual total cost of JPY 2000 to consumers and companies.⁶¹

*[c]    Potential Challenges*

Even with this new program, there are still some challenges ahead. For instance, for PV, with the introduction of the FIT, solar power is expected to expand rapidly, but other energy sources will be limited to the growth of small-scale plants because of geography, technological limitations, time required for environmental impact assessments, and weak grid connections that make it difficult to alternate between sources of renewable power.⁶² In a recent interview, Tetsuro Nagata, head of the Japan Windpower Association, said wind power is concentrated in Hokkaido and the Tohoku region and that unless grid connections to deliver the power from remote areas are strengthened, it will be difficult to expand, even with the new tariff. For geothermal energy, due to strict environmental regulations and construction standards, 10 years may be required for a geothermal plant to begin operation.⁶³

**[B]    South Korea**

Promotion of renewable electricity in Korea can be divided into three main phases. During Phase I, an early law was enacted – the New and Renewable Energy Development and Promotion Act of 1987 – to promote the overall development of renewable

---

59. METI, Feed in Tariff Scheme for Renewable Energy-Launched on 1 Jul. 2012, at p. 8, available at: http://www.meti.go.jp/english/policy/energy_environment/renewable/pdf/summary2012 09.pdf (visited on 1 Sep. 2012).
60. Feed-in tariff era gets under way, Monday, 2 Jul. 2012 available at: http://www.japantimes. co.jp/text/nn20120702a2.html (visited on 1 Sep. 2012).
61. Japanese Times, Japan added Environmental Electricity Capacity equivalent to two Nuclear Power Plants, 2012/07/02, available at: http://zh.cn.nikkei.com/industry/ienvironment/2838-20120702.html (visited on 1 Sep. 2012).
62. ERIC JOHNSTON, Feed-in tariff has solar advocates sky high, Saturday, 30 Jun. 2012, available at:http://www.japantimes.co.jp/text/nn20120630f1.html (visited on 1 Sep. 2012).
63. ERIC JOHNSTON, Feed-in tariff has solar advocates sky high, Saturday, 30 Jun. 2012, available at:http://www.japantimes.co.jp/text/nn20120630f1.html (visited on 1 Sep. 2012).

energy.[64] Then, in the 2000s, there was a series of legal actions giving rise to a 'feed-in tariff scheme.' However, this scheme faced some political challenges and was replaced by an RPS scheme in 2010.[65]

### [1] Phase I. Soft Law in the 1990s

During this first stage, the priority was RD&D, rather than market applications. A substantial amount of *RD&D funding and projects* were carried out during this period; between 1988 and 1998, public money was invested in ~300 projects in 11 research areas, including PV, bio-energy, waste energy, wind power, solar, ocean, geothermal, hydrogen, and small hydro projects. Fuel cells and clean coal use were also funded.[66]

Since 2000, a number of government policies, such as the 2001 *Alternative Energy RD&D Basic Plan*, *Second Basic Plan for National Energy*, and *Basic Plan for NRE Technology Development and Dissemination*,[67] sought to promote development and market deployment of renewable electricity, leading to the introduction of the FIT scheme and several other promotion schemes.

### [2] Phase II: 2000s

#### [a] Feed-In Tariffs

Because the previous versions of the NRE law did not provide more effective measures for market deployment, the government began to adopt a series of legal programs to establish a FIT scheme under the *2001 Electricity Act*,[68] *2002 Revision of NRE Law*,[69] '*Utilization of New and Renewable Energy for Generation of Electric Power Standard Tariff Guide*'[70] and other schemes.[71]

*Fifteen- and twenty-year tariffs.* Originally, the standard FIT for different types of renewable electricity was applied for 15 years [PV, wind, small hydro, landfill gas (LFG), tidal, waste].[72] The rate schedule before and after 10 October 2006, please see Tables 9.7 and 9.8.

---

64. IEA, ENERGY POLICIES OF IEA COUNTRIES: SOUTH KOREA 2002 48 (2002).
65. ACT ON THE PROMOTION OF THE DEVELOPMENT, USE, AND DIFFUSION OF NEW AND RENEWABLE ENERGY, Amended by Act No. 10253, 12 Apr. 2010.
66. IEA, *supra* note 64, at 48.
67. IEA, ENERGY POLICIES OF IEA COUNTRIES: SOUTH KOREA 2006 71 (2006).
68. Article 49 of Electric Business Law.
69. Article 17 of the Development, Use, Supply and Promotion of New and Renewable Energy Law.
70. 'Utilization of New and Renewable Energy for Generation of Electric Power Standard Tariff Guide' (Ministry of Commerce, Industry and Energy Official Notice No. 2002-108, 29 May 2002, 26 Sep. 2002 amendment notification, 9 Oct. 2003, 19 Oct. 2004 amendment).
71. For example, Arts 4 and 10 of Other Energy Support Business Operation Outline (MCIE Notice No. 2002-034).
72. Article 7 (Term of application of the base price): 'The base price as decided by this guide is applicable from the start date of supply of new and renewable energy to Korean electricity supply channels and the base price apply period will not exceed 15 years from the start date of commercial operation.'

*Table 9.7  Applied Standard and Base Price for Corporations Purchasing Sources of Electricity (before 10 October 2006)*

| Target Source of Electric Power | Applied Facility Volume Standard | Base Price (KRW/kWh) | | Notes |
|---|---|---|---|---|
| Solar | ≥3 kW | 716.40 | | |
| Wind | ≥10 kW | 107.66 | | |
| Tidal (breakwater) | ≥50 MW | 62.81 | | |
| Small hydraulic power | ≤3 MW | 73.69 | | |
| Reclaimed gas | ≤50 MW | <20 MW | 65.20 | Fossil fuel investment: <30% |
| | | 20–50 MW | 61.80 | |
| Waste incineration (RDF included) | ≤20 MW | SMP + CP | | |

*Notes*: Application of the base price is limited to <30% of the government subsidy funding. The standard for fossil fuel investment is measured per month and is the percentage of energy used to produce electric power that originates from fossil fuel energy.

*Table 9.8  Applied Standard and Base Price for Corporations Purchasing Sources of Electricity (Post-11 October 2006)*

| Source | Applied Facility Volume Standard | Classification | Base Price (KRW/kWh) | | Notes |
|---|---|---|---|---|---|
| | | | Fixed | Variable | |
| Solar | ≥3 kW | ≥30 kW | 677.38 | – | Decrease 4% (after three yrs) |
| | | <30 kW | 711.25 | – | |
| Wind | ≥10 kW | – | 107.29 | | Decrease 2% (after three yrs) |

---

*See also*, IEA, Turning a Liability into an Asset: the Importance of Policy in Fostering Landfill Gas Use Worldwide 16 (2009); OSEC, Market Study: Green Technology in Korea 12 (2010); PV Tech, South Korea, available at: http://www.pv-tech.org/tariff_watch/south_korea (visited on 1 Sep. 2012).

| Source | | | Applied Facility Volume Standard | Classification | | Base Price (KRW/kWh) | | Notes |
|---|---|---|---|---|---|---|---|---|
| | | | | | | Fixed | Variable | |
| Hydraulic | | | ≤5 MW | General/ majority | ≥1 MW | 86.04 | SMP + 15 | |
| | | | | | <1 MW | 94.64 | SMP + 20 | |
| | | | | Other | ≥1 MW | 66.18 | SMP + 5 | |
| | | | | | <1 MW | 72.80 | SMP + 10 | |
| Waste incineration (RDF included) | | | ≤20 MW | - | | – | SMP + 5 | Fossil fuel investment: <30% |
| Bio-energy | LFG | | ≤50 MW | ≥20 MW | | 68.07 | SMP + 5 | |
| | | | | <20 MW | | 74.99 | SMP + 10 | |
| | Bio-gas | | ≤50 MW | ≥150 kW | | 72.73 | SMP + 10 | |
| | | | | <150 kW | | 85.71 | SMP + 15 | |
| | Biomass | | ≤50 MW | Wood biomass | | 68.99 | SMP + 5 | |
| Ocean | Tidal | | ≥50 MW | Max. tidal range ≥8.5 m | With seawall | 62.81 | | |
| | | | | | Without seawall | 76.63 | | |
| | | | | Max. tidal range <8.5 m | With seawall | 75.59 | | |
| | | | | | Without seawall | 90.50 | | |
| Fuel cell | | | ≥200 kW | Use of bio-gas | | 234.53 | | Decrease 3% (after two yrs) |
| | | | | Use of other Fuels | | 282.54 | | |

*Source:* http://www.osec.ch/en/filefield-private/files/6561/field_blog_public_files/7928, p. 12.

*Notes:* Application of the base price is limited to <30% of the government subsidy funding.

The standard for fossil fuel investment is measured per month and is the percentage of energy used to produce electric power originating from fossil fuels.

A facility's capacity for solar powered electricity is the rated module capacity.

The majority of hydraulic power is focused on hydraulic-powered electricity generation; the rest is focused on supplementary hydraulic-powered electricity generation

SMP: System Marginal Price (cost of the most expensive generating unit included in the price setting schedule of the Korea Power Exchange at a trading day; 2009 average: KRW 105.04/kWh)

However, since October 2008, PV installed between 1 October 2008 and 31 December 2009 may choose between a lower rate for 20 years and a higher rate for 15 years. The rate schedule for PV from 1 October 2008 is provided as Table 9.9.

Table 9.9   The Rate Schedule for PV (from 1 October 2008) Unit: Korean Yuan

| The Year | Type | Purchase Period (Yr.) | 30 kW≤ | > 30 kW ≤200 kW | > 200 kW ≤1 MW | > 1 MW ≤3 MW | > 3 MW |
|---|---|---|---|---|---|---|---|
| 2008-2009 | Ground type | 15 | 484.52 | 462.69 | 436.5 | 414.68 | 349.2 |
| | | 20 | 439.56 | 419.76 | 396 | 376.20 | 316.80 |
| | Rooftop type | 15 | 532.97 | 508.96 | 480.15 | - | - |
| | | 20 | 483.52 | 461.74 | 435.60 | - | - |

For those who receive the confirmation of the setting up between October 2008 and December 2009, they can choose the purchase period. See www. Knerec.or.kr.

Renewable electricity generators bid into the Korea Power Exchange (KPX). The government compensates eligible renewable energy generators for any shortfall between the pool price and the preset feed-in tariff. The first- and second-phase rate schedules are shown in Tables 9.10 and 9.11. Because of the PV boom, South Korea introduced a prompt tariff cut, resulting in rates for 2010[73] (see Table 9.10) and 2011 that are much lower than the original rate.

Table 9.10   PV Feed-In Traiffs Since 2010 in KRW/kWh

| | Runtime | < 30 kW | 30–200 kW | 200 kW–1 MW | 1-3 MW | > 3 MW |
|---|---|---|---|---|---|---|
| General | 15 years | 566.95 | 541.42 | 510.77 | 485.23 | 480.62 |
| | 20 years | 514.34 | 491.17 | 463.37 | 440.20 | 370.70 |
| BIPV* | 15 years | 606.64 | 579.32 | 546.52 | – | – |
| | 20 years | 550.34 | 525.55 | 495.81 | – | – |

Source: KEMCO

*Building Integrated Photovoltaic Modules.

*Tariff degression.* The original plan shown in Tables 9.10 and 9.11 is effective until the end of 2020. To respond to changes over time in the cost of renewable electricity, the

---

73. OSEC, *supra* note 72, at 11.

rate schedule also includes a degression table. PV and wind rates began to degress after three years of unchanged rates, while fuel cell rates degressed after two years. The complete tariff degression, please see Table 9.11.

Table 9.11  Tariff Degression in South Korea (KRW/kWh)

| Year of Commercial Operation | Solar Energy | | Wind | Fuel Cells | |
|---|---|---|---|---|---|
| | > 30 kW | < 30 kW | | Bio-Gas | OtherFuels |
| 2006 | 677.38 | 711.25 | 107.29 | 234.53 | 282.54 |
| 2007 | 677.38 | 711.25 | 107.29 | 234.53 | 282.54 |
| 2008 | 677.38 | 711.25 | 107.29 | 227.49 | 274.06 |
| 2009 | 650.28 | 682.80 | 105.14 | 220.67 | 265.84 |
| 2010 | 624.27 | 655.49 | 103.04 | 214.05 | 257.87 |
| 2011 | 599.30 | 629.27 | 100.98 | 207.63 | 250.13 |
| 2012 | 575.33 | 604.10 | 98.96 | 201.40 | 242.63 |
| 2013 | 552.32 | 579.93 | 96.98 | 195.36 | 235.35 |
| 2014 | 530.22 | 556.74 | 95.04 | 189.50 | 228.29 |
| 2015 | 509.02 | 534.47 | 93.14 | 183.81 | 221.44 |
| 2016 | 488.65 | 513.09 | 91.28 | 178.30 | 214.80 |
| 2017 | 469.11 | 492.56 | 89.45 | 172.95 | 208.35 |
| 2018 | 450.34 | 472.86 | 87.66 | 167.76 | 202.10 |
| 2019 | 432.33 | 453.95 | 85.91 | 162.73 | 196.04 |
| 2020 | 415.04 | 435.79 | 84.19 | 157.85 | 190.16 |

Note: The degressed rate is applied beginning on 11 October of each year.

Cap. In 2002, the government set an upper limit on support for renewables of 250 MW for wind and 20 MW for solar. The guaranteed feed-in tariff is granted on a first-come, first-served basis up to the limit.[74] In the 2006 Guideline, the caps for solar, wind, and fuel cells were set to 100, 1,000, and 50 MW, respectively, of accumulated equipment volume.[75] The *PV total cap* was raised to 500 MW in 2008 to respond to the rapid development of PV,[76] and the other caps were set to 340 and 400 MW, respectively. In addition, there are specific *annual caps* for PV and fuel cells for 2009, 2010, and 2011. Due to the proliferation of PV, the government limited support for the development of

---

74. IEA, *supra* note 67, at 75.
75. Clause 3 (Limits to applied volume of the base price) Out of the target power sources in Table 9.1 to which the base price applies, the base price apply target for solar, wind and fuel-cell is set to, respectively, 100 MW, 1,000 MW, 50 MW of accumulated equipment volume. *See* Ministry of Commerce, Industry and Energy Notice No. 2006-89, 30 Aug. 2006, Minister of Commerce, Industry and Energy, 'Utilization of New and Renewable Energy for Generation of Electric Power Standard Tariff Guide', available at: www.climatelaw.org/laws/korea/refit (visited on 1 Sep. 2012).
76. Basant Agrawal, Gopal Nath Tiwari, Building Integrated Photovoltaic Thermal Systems: For Sustainable Developments 188 (2010).

renewable electricity to a PV cap of 162 MW, while the 2008 caps for other renewable electricity sources of 340 and 400 MW were reduced to 98 and 132 MW, respectively. (See Table 9.12.)

*Table 9.12   Annual Caps for PV and Fuel Cells*

|  | PV | Fuel Cells |
|---|---|---|
| 2009 | 50 MW | 12 MW |
| 2010 | 70 MW | 14 MW |
| 2011 | 80 MW | 16 MW |

*Source*: http://www.osec.ch/en/filefield-private/files/6561/field_blog_public_files/7928, p. 11.

*Cost-sharing mechanism.* The South Korean government has covered the costs of the feed-in tariff out of its own budget, a move that quickly became a strain on its finances.[77] As the funding is not covered by the Korean consumers,[78] this became substantial burden on the South Korean government, and also on the 'taxpayer's money.'[79]

*Evaluation.* Generally speaking, the PV boom was a turning point in the South Korean FIT scheme. The original three-year unchanged FIT and the inadequate cost-sharing scheme were significant flaws that resulted in a heavy governmental burden, leading to the PV boom and the dramatic FIT rate cut for PV in 2008.[80] In October 2008, the South Korean government slashed the feed-in tariff by 8%–30%, and as a result, the Korean PV market installed only 10 MW between October 2008 and March 2009.[81] In addition, the government also established a new requirement that PV projects must be completed within three months.[82] The proliferation of PV ultimately led to the collapse of the FIT scheme.

---

77. Jane Burgermeister, South Korea Taps Germany To Help Grow Its Solar Industry, available at: http://www.renewableenergyworld.com/rea/news/article/2009/04/south-korea-looks-to-germany-to-help-grow-its-solar-industry (visited on 1 Sep. 2012).
78. Article 3 (Terminology): '"Government Support Fund" refers to the *rationalization* laws regarding use of energy when establishing new and renewable electric energy facilities; to Development, Use, Supply and Promotion of New/Renewable Energy Law; to the Electric Business Law; to Government subsidy; and to funds and national treasury grants.
    "Support ratio" refers to the proportion of overall funds used for development and installation of new and renewable energies that are provided by the Government.'
    *See also*, OSEC, *supra* note 72, at 11.
79. Miguel Mendonça et al., Powering the Green Economy: the Feed-in Tariff Handbook 61 (2009).
80. Jane Burgermeister, South Korea Taps Germany To Help Grow Its Solar Industry, available at: http://www.renewableenergyworld.com/rea/news/article/2009/04/south-korea-looks-to-germany-to-help-grow-its-solar-industry (visited on 1 Sep. 2012).
81. PV Tech, South Korea, 22 Dec. 2011, available at: http://www.pv-tech.org/tariff_watch/south_korea (visited on 1 Sep. 2012).
82. OSEC, *supra* note 72, at 11.

### [b] Other Supplementary Schemes

Along with the FIT, several other incentives are also provided in South Korea.

#### [i] Investment Subsidies and Demonstration Project Subsidy

The government has designated hydrogen fuel cells, PV, and wind as areas that will receive the largest share of government support – 70%. Between 1996 and 2004, the government provided KRW 234.8 billion in direct support for construction and operation of renewable power plants.[83]

The government provides funds required for construction and operation of facilities for new and renewable energy, such as solar thermal energy and PV energy. The percentage of the government's support ranges up to *100%* of the required funds. Support terms and conditions include (1) a *2.0%* annual percentage rate (APR), (2) a *10-year* redemption period, and (3) amortization after a *5-year* deferment period.

For example, to promote solar power, both to reduce domestic fossil fuel use and develop a long-term export market, the government is supporting the construction of 100,000 homes that rely on solar PV power for some of their power needs. To meet this goal, the government has provided subsidies totaling KRW 72.8 billion to 332 projects between 2001 and 2004, with a total established capacity of 837 kW. This project is expected to save 285 toe/year of energy, equivalent to about KRW 120 million in oil imports per year.[84]

#### [ii] Taxes

For companies making investments in new and renewable power generation facilities, a one-time deduction of 10% of their investment in R&D can be taken from the builder's individual income or corporate tax.[85]

#### [iii] Customs Duties

Sixty-five percent of the customs levied on 26 different items in 4 categories (solar thermal energy, PV energy, wind power, and fuel cell energy) can be deducted.[86]

---

83. IEA, *supra* note 67, at 76.
84. IEA, *supra* note 67, at 76.
85. Eanst & Young, 2011 Asia-Pacific R&D incentives 7 (2011).
86. IEA, *supra* note 67, at 76.

## Chapter 9: Transformation of FIT in Japan, South Korea, and Taiwan  §9.02[B]

[iv]   Loans

The government provides low-interest loans to installers of NRE equipment. For example, in 2002, KRW 23.4 billion[87] was provided at a floating annual interest rate of 4.75%, with a term of repayment of five years in equal monthly installments and a three-year grace period.

In 2006, loan funding was focused mainly on solar PV and biomass.[88] Financial assistance included low-interest loans (5.5% with a three year grace period and five years repayment period) for companies that installed renewable energy technologies, up to 80% of the total installation cost.[89]

Table 9.13 demonstrates that loans are the tool most frequently used by the South Korean government in facilitating development of renewable energy over the past few years.

*Table 9.13   Loans and Other Financial Support for Development of Renewable Energy in South Korea*

|  | (Unit: Million Won) | | | | | |
|---|---|---|---|---|---|---|
|  | 1988~1998 | 1999 | 2000 | 2001 | 2002 | Total |
| R&D | 137,832 | 18,201 | 18,192 | 29,125 | 32,745 | 236,095 |
| Demonstration | 10,817 | 5,568 | 9,130 | 23,231 | 23,539 | 72,285 |
| Loans | 247,741 | 14,322 | 11,707 | 8,498 | 23,400 | 305,668 |
| Total | 396,390 | 38,091 | 39,029 | 60,854 | 79,684 | 614,048 |
| **Estimates for the Year 2002** | | | | | | |

*Source:* Korea's Policy Direction for R&D and Dissemination of New & Renewable Energy, November 2002, Korea Energy Management Corporation.

The funding sources for these low-interest loans are diverse and include:[90]

- Green Growth Corporate Loan.
- Korea Credit Guarantee Fund.
- Kibo Technology Fund.
- Solar Energy Development Facilities Fund Loan.

---

87. Korea's Policy Direction for R&D and Dissemination of New & Renewable Energy, November, 2002 Korea Energy Management Corporation.
88. IEA, *supra* note 67, at 75.
89. IEA, FINANCIAL ASSISTANCE TO ENERGY EFFICIENCY INVESTMENTS of South Korea, available at: http://www.ieadsm.org/Files/Exco%20File%20Library/Country%20Publications/FINANCE.doc (visited on 1 Sep. 2012).
90. Korea Times, IBK Gears Up to Bolster Green Growth Industries, 10 Jan. 2009, available at: http://www.koreatimes.co.kr/www/news/biz/2011/03/283_52835.html (visited on 1 Sep. 2012).

- Green Growth Deposit.
- Export-Import Bank of Korea Carbon Fund.

[v]     Voluntary Renewable Portfolio Agreement

The South Korean government has signed a renewable portfolio agreement (RPA) with nine energy-related public organizations, including Korea Electric Power Corporation (KEPCO), six power companies, Korea District Heating Corporation (KDHC), and Korea Water Resources Corporation (K-Water). This agreement was signed on 10 July 2009 to elicit voluntary investment in renewable energy. Under the RPA, these public companies voluntarily pledged to take a leading role in using renewable energy, supporting R&D activities, and staging public relations campaigns.[91] Under a 2005 RPA, KEPCO invested USD 740 million in building renewable energy facilities with a capacity of 332 MW over three years. In addition, the company is funding R&D to replace fossil fuels with green energy.[92]

[vi]     Mandatory Renewable Standards for Public Buildings

In 2002, to promote the dissemination of new and renewable energy, the government passed legislation (*2002 Revision of NRE Development and Promotion Act*) requiring that all newly built public buildings (including federal and local government buildings) with over 3,000 m² gross area allocate > 5% of their construction costs to establishment of new and renewable energy facilities. In 2004, the government provided KRW 5 billion to 10 model businesses, the energy production capacity of which is estimated to be 805 toe. New markets valued at KRW 100–200 billion annually are expected to be generated by this program.[93]

*[3]     Phase III. 2012: RPS and Mandatory Capacity Installation of PV*

[a]     RPS

Due to the proliferation of PV in 2008, South Korea began to consider changing its FIT scheme to an RPS scheme. The RPS law, *Act on the Promotion of the Development, Use, and Diffusion of New and Renewable Energy*, Amended by Act No. 10253, 12 April 2010, was adopted in 2010 and became effective in January 2012. The official rationale for this policy remains the pursuit of energy policy goals, but it is unclear why a more

---

91. J.W. Chang et al., Overview of renewable energy development in South Korea and the economic analysis of grid interconnection for large scale offshore wind farm, at page 4, available at: http://www.kemco.or.kr/new_eng/pg02/pg02040705.asp (visited on 1 Sep. 2012).
92. PWC, Carbon Disclosure Project Report 2008 Global 500, at page 75, available at: https://www.cdp.net/CDPResults/67_329_143_CDP%20Global%20500%20Report%202008.pdf (visited on 1 Sep. 2012).
93. IEA, *supra* note 67, at 76–77.

effective FIT could not achieve the same goal. It is quite ironic that President Lee Myung-bak promoted such a scheme in his *Green Energy Strategy* of September 2008 and the *Third Renewable Energy Plan* of December 2008 under his ambitious plan to pursue low carbon green growth. The FIT scheme and RPA were completely abolished at the end of 2011.

Under the new RPS scheme, electricity generation companies (electricity suppliers)[94] are required to provide a certain percentage of electricity from renewable electricity sources.[95] Electricity generation companies include all those with capacities > 500 MW,[96] six large-capacity companies (with installation capacity much larger than 500 MW),[97] and seven smaller companies.[98] Not all renewable electricity qualifies; the Act only applies to those renewable energy sources whose technologies and costs can

---

94. It is really strange that Korea consider these two companies are the same idea. *See* e.g., Min-Cheol, Kang, Renewable Portfolio Standard in the Republic of Korea, 3 May 2011, available at: http://www.oav.de/uploads/tx_ttnews/kr_110503_kemco_rps.pdf (visited on 1 Sep. 2012).
    Yet in Europe, electricity generation companies is different from electricity supply company, e.g., in Sweden, only electricity suppliers, instead of electricity generators, are subject to the RPS scheme. *See* Sweden, Quota system, available at: http://www.res-legal.eu/search-by-country/sweden/single/s/res-e/t/promotion/aid/quota-system-1/lastp/199/.
95. Article 12-5 (Establishing Responsibility for Supply of New and Renewable Energy, etc.):
    '(1) Where the Minister of Knowledge Economy deems it necessary to promote use and diffusion of new and renewable energy and facilitate the new and renewable energy industry, he/she may require an entity prescribed by Presidential Decree from among those falling under any of the following subparagraphs (hereinafter referred to as "entity with responsibility to supply") to supply a certain or larger amount of power generation with new and renewable energy:

    1. An operator of an electric power generation business under Art. 2 of the Electric Business Act;
    2. An entity deemed to have acquired a license to run an electric power generation business under Art. 7(1) of the Electricity Business Act in accordance with Arts 9 and 48 of the Integrated Energy Supply Act; and
    3. Public institutions;

    (2) The total quantity of power that an entity with responsibility to supply should mandatorily supply by new and renewable energy under paragraph (1) (hereinafter referred to as "mandatory supply quantity") shall be annually prescribed by Presidential Decree within 10% of the quantity of total power generation. With regard to the kinds of new and renewable energy that require balanced use and diffusion, part of the total mandatory supply quantity may be supplied using the appropriate kind of new and renewable energy as prescribed by Presidential Decree.
    (3) The mandatory supply quantity of an entity with responsibility to supply shall be prescribed and publicly announced by the Minister of Knowledge Economy for each responsible entity after hearing the opinions of the responsible entities. The Minister of Knowledge Economy shall consider the total amount of power generation of responsible entities, sources of power generation, etc.'
96. REEGLE, South Korea 2012, available at: http://www.reegle.info/policy-and-regulatory-overviews/KR (visited on 1 Sep. 2012).
97. Korea Hydro Nuclear Power Co., Ltd. (KHNP) – 2,449 GWh; South East Power Co., Ltd (SEP) – 1,080 GWh; Korea West Power Co., Ltd. (KWP) – 976 GWh; Korea Southern Power Co., Ltd., (KOSPO) – 965 GWh; Korea Midland Power Co., Ltd (KOMIPO) – 953 GWh; Korea East West Power Co., Ltd. (KEWP) – 875 GWh.
98. Korea District Heating Corporation (KDHC);K-Water Corporation; SK E&S (SK E&S is under M&A process. Used to be K-Power);GS EPS; GS Power; POSCO Power; MPC (Yulchon).

be verified and commercialized.[99] Under this definition, only PV, wind, hydropower, tidal, bio-gas, landfill gas (LFG), biomass, fuel cells, integrated gasification combined cycle (IGCC), waste, and refuse-derived fuel (RDF) are included.

The compulsory RPS rate is set at 2% in 2012, then increases gradually to 10% in 2022.[100] (See Table 9.14) With respect to the rate, it is important to note that not all renewable energy output is counted in the same manner. It was anticipated that without an adjustment factor, the RPS scheme would only encourage low-cost renewable energy generation. Thus, to promote diverse types of renewable energy simultaneously, a 'multiplier' is included that takes into account five factors (Table 9.15): (1) economics, (2) environment, (3) potential, (4) industrial effects, and (5) consistency with the policy direction. For example, because offshore wind is more expensive, a multiplier of >1.5 is applied. For waste, due to environmental concerns and its relatively low-cost, its contribution to renewable energy goals is discounted by 50%.

Table 9.14  Compulsory RPS Rate in South Korea

| Year | 2012 | 2013 | 2014 | 2015 | 2016 | 2017 | 2018 | 2019 | 2020 | 2021 | 2022 |
|---|---|---|---|---|---|---|---|---|---|---|---|
| Compulsory Ratio (%) | 2.0 | 2.5 | 3.0 | 3.5 | 4.0 | 5.0 | 6.0 | 7.0 | 8.0 | 9.0 | 10.0 |

Table 9.15  RPS Rate Multipliers in South Korea

| | | Eligible Resources | | |
|---|---|---|---|---|
| Grouping | Multiplier | Installation Type | Land Type | Capacity |
| Solar Energy | 0.7 | In case of not use 'Building and Existing Facilities' | 5 lands (Rice field, Dry field, Orchard, Pasture, Forest land) | |
| | 1.0 | | Other | Excess 30 kW |
| | 1.2 | | | Under 30 kW |
| | 1.5 | Building and Existing Facilities | | |
| General Renewable | 0.25 | • IGCC | | |
| | 0.5 | • Waste | • LFG (Land fill gas) | |
| | 1.0 | • Hydro | • Wind (on-shore) | |
| | | • Bio-gas | • Bio-mass | |
| | | • Tidal I (construction under having its tide embankment) | | |

---

99. MKE, Renewable Portfolio Standard in the Republic of Korea, available at: http://www.bc bioenergy.com/wp-content/uploads/2012/03/2-Korean-Policy-Environment-for-Renewable-Fuels-Tae-Jun-Choi.pdf (visited on 1 Sep. 2012).
100. MKE, Renewable Portfolio Standard in the Republic of Korea, available at: http://www.bc bioenergy.com/wp-content/uploads/2012/03/2-Korean-Policy-Environment-for-Renewable-Fuels-Tae-Jun-Choi.pdf (visited on 1 Sep. 2012).

|  | 1.5 | • Off-shore Wind (connecting point length is under 5 km) |
|---|---|---|
|  | 2.0 | • Off-shore Wind (connecting point length is over 5 km)<br>• Tidal II (newly construction its tide embankment)<br>• Fuel-cell |

*Source:* MKE, Renewable Portfolio Standard in the Republic of Korea, available at: http://www.bcbioenergy.com/wp-content/uploads/2012/03/2-Korean-Policy-Environment-for-Renewable-Fuels-Tae-Jun-Choi.pdf (visited on 1 September 2012).

KEMCO will provide primary administration of this RPS.[101] The requirement can be fulfilled in three ways: (1) generation by the company, (2) buying electricity from other companies, and (3) acquiring sufficient certificates.[102]

[b]  *Mandatory Capacity Installation of PV*

For solar energy, the South Korean government adopted two protective measures. First, as PV is very competitive among Korean industries, there is a dedicated set of multipliers designed specifically for PV; building-integrated PV (BIPV) is relatively expensive, and larger PV systems are less expensive than smaller ones. Therefore, different multipliers apply. There is also a minimum requirement for PV electricity output and installation on an annual basis to provide further protection to PV (Table 9.16). However, it is unclear how this requirement should be distributed among electricity users, i.e., whether all should divide this requirement proportionately.

*Table 9.16  Minimum Requirement for PV Electricity Output Each Year*

| Year | 2012 | 2013 | 2014 | 2015 | 2016~ |
|---|---|---|---|---|---|
| Obligation | 276 GWh | 591 GWh | 907 GWh | 1,235 GWh | 1,577 GWh |
| Year | 2012 | 2013 | 2014 | 2015 | 2016~ |
| Capacity | 220 MW | 450 MW | 690 MW | 1040 MW | 1200 MW |

[c]  *Other Measures*

Other measures are provided in the 2010 revision of the Law for support of renewable electricity development:

---

101. KEMCO, Certificate of NRE systems, available at:http://www.kemco.or.kr/new_eng/pg02/pg02040703.asp (visited on 1 Sep. 2012).
102. Article 12-5: '5) Responsible entities may meet the mandatory supply quantity by purchasing supply certification of new and renewable energy under Article 12-7.'

- Basic plan[103] and yearly implementation plan.[104]
- Establishment of a special task force.[105]
- Development of financial tools,[106] such as (1) all renewable energy technologies receive a *5% tax credit*,[107] (2) in 2009, *import duties* were halved on all components/equipment used in renewable energy power plants, (3) *government subsidies* are available *to local governments* of up to 60% for installation of renewable facilities, as well as *low-interest loans (5.5%–7.5%)* for renewable energy projects with a 5-year grace period followed by a 10-year repayment period.[108]
- Establishment of a research centre responsible for renewable energy R&D.[109]

[C]    Taiwan

*[1]    Phase I: Before Drafting Renewable Energy Bill in 2002*

Before 2002, the main form of renewable electricity in Taiwan's electricity mix, i.e., hydro power, was not developed under any incentive and legislative scheme. Rather, it was ambitiously developed during the Japanese colonization of Taiwan; Tai-Power has continued to follow governmental policy in exploiting hydropower since the KMT's moving to Taiwan after World War II, in the 1940s. Hydropower has been rapidly developing during this stage and large-scale hydropower is heavily exploited, so that room for only small hydropower plants remains in the future. In addition to hydropower, there is also a geothermal electricity site developed voluntarily by the CPC.

---

103. Article 5 (Establishment of Basic Plan).
104. Article 6 (Yearly Implementation Plan).
105. Article 8 (New and Renewable Energy Policy Council):
    '(1) For the purpose of deliberating on important matters concerning the technological development, use, and diffusion of new and renewable energy, the New and Renewable Energy Policy Council (hereinafter referred to as "Council") shall be established within the Ministry of Knowledge Economy.'
106. Article 29 (Financial Measures, etc.): 'The State shall take necessary measures, such as providing financial or tax support, where it is necessary, to an entity that is recommended pursuant to Article 12 or that must comply with duties, an entity engaged in the technological development, use, and diffusion of new and renewable energy, or an entity that has obtained certification of facilities under Article 13'.
107. REEEP, Policy DB Details: South Korea (2012), available at: http://www.reeep.org/index.php?id=9353&text=policy-db&special=viewitem&cid=151 (visited on 1 Sep. 2012).
108. Young Il Choung, Quick Look: Renewable Energy Development in South Korea, available at: http://www.renewableenergyworld.com/rea/news/article/2010/12/quick-look-renewable-energy-development-in-south-korea (visited on 1 Sep. 2012).
109. Article 31 (New and Renewable Energy Center): '(1) The Minister of Knowledge Economy may establish a new and renewable energy center (hereinafter referred to as "Center") under an energy agency prescribed by Presidential Decree, in order to professionally and efficiently promote the use and diffusion of new and renewable energy, and may have the Center perform the projects falling under each of the following subparagraphs.'

The development of biomass electricity was influenced by incentives. Renewable electricity from municipal waste was encouraged by the Energy Management Act[110] and CHP Ordinance.[111] Tai-Power is obliged to buy the electricity from municipal waste incinerators at a preferential price. A similar regime is also applied to bio-gas from the landfill site.[112] As there is a net-metering scheme in place, these forms of renewable electricity are developing rapidly.

Compared with the aforementioned early development of renewable electricity, the development of PV and wind power came quite late. Since 2000, they have been promoted by investment and a demonstration subsidy scheme, respectively.[113] However, since there is no feed-in tariff scheme in place, they are not promoted very well.

*[2]   Phase II: 2003 Restricted Version of FIT: Transition Period between 2002 and 2009*

As noted above, a similar scheme was used to promote electricity from bio-gas[114] and municipal waste incinerators.[115]

Regarding the development of a stationary biopower system, currently a total installed capacity of 772 MW for power generation has been achieved through biomass energy utilization, which is primarily based on MSW incineration generation, bio-gas power generation, and the utilization of agricultural and industrial wastes. There are currently 25 operational large incineration plants, with a total installed capacity of 622.5 MW. The bio-gas electricity generation systems with a total installed capacity of 24.5 MW include four landfill sites. As for the utilization of agricultural and industrial wastes – bagasse, paper rejects, waste plastics and rubber, and rice husks – these densified RDF are used for generation with an installed capacity of 125 MW.[116]

Even though Taiwan is located in the Pacific Rim, there are limited suitable sites to develop large-scale geothermal electricity. The Tatun volcanic area has the largest geothermal resource in Taiwan. Its development, however, was halted due to severe corrosion resulting from acid fluids and constraints on land accessibility under the National Park Act. Other than the Tatun volcanic area, Ching-Shui of Ilan County is currently considered more appropriate for geothermal development. In August 2005, the Bureau of Energy in the MOEA announced the 'Ordinance on Exploration Subsidy Program for Geothermal Power Generation and Demonstration.'[117] Up to the present, the County Government of Ilan has been awarded a project to implement 'The Study of

---

110. Article 10 of Energy Management Act.
111. CHP Ordinance (first enacted on 15 Jul. 1988).
112. Biogas Feed-in Tariff Ordinance, available at: http://ivy5.epa.gov.tw/epalaw/docfile/049 146.doc.
113. Ordinance on Demonstration Investment Subsidy of Wind Power Plants (22 Mar. 2000); Ordinance on Demonstration Investment Subsidy of PV Plants (31 May 2000).
114. Biogas Electricity Feed-in Tariff Ordinance (enacted on 22 Jan. 2003; last revised 13 Oct. 2009).
115. Article 10 of Energy Management Act; CHP Ordinance I (first enacted on 15 Jul.1988); CHP Ordinance II (enacted on 4 Sep. 2002; last revised 27 Dec. 2010).
116. http://www.moeaboe.gov.tw/About/webpage/book_en1/page3.htm.
117. Ordinance on Exploration Subsidy for Geothermal Electricity Demonstration System.

Geothermal Resource and Power Generation Potential in Ching-Shui, Ilan.' The MOEABOE will continue to help Ilan County in establishing a demonstrative showcase for the integration of power generation and recreation. In addition, the Geothermal Exploration Subsidy Program granted an application by Taitung County Government in 2007, for help in conducting geothermal exploration and an evaluation of power generation potential in the Jinlun area.[118]

The wind power promotion scheme underwent dramatic change during this phase. As the investment subsidy was not sufficient to encourage the deployment of wind power, this scheme was abolished in 2003. A 'temporary' feed-in tariff scheme was developed by Tai-Power. Tai-Power voluntarily enacted its 'feed-in tariff Ordinance' in 2003.[119] According to Article 6 of this Ordinance, the buyback tariff for all renewable electricity is set at NTD 2/kWh. This scheme was intended to promote many types of renewable electricity (PV, marine energy, wind, geothermal, small hydropower, biomass [excluding municipal waste incinerators]).[120] Yet, due to the fixed NTD 2/kWh incentive, mainly wind power plants were introduced at this stage. The aforementioned Infravest private company developed many wind turbines during this period. According to the statistics in 2007, Tai-Power and Infravest were the two major players in building wind turbines. Taiwan has completed 155 sets of wind turbines with a total capacity of 281.6 MW, built by Tai-Power and some private sector organizations. Assuming that 1 kW of the installed capacity produces on average 2,700 kWh per year, wind power can generate 760 million kWh annually, providing enough electricity for 190 thousand households (a family of 4). Moreover, projects under construction, being approved, and being planned will offer a total installed capacity of 467.8 MW (equivalent to 230 sets of wind turbines).[121] Please see the Table 9.17 below for the complete list of developers.

*Table 9.17*

|   | Wind Farm | Developer | Number of Wind Turbines | Capacity (MW) |
|---|---|---|---|---|
| 1 | Penghu JhongtunFirst-phase | TPC | 4 | 2.4 |
| 2 | Shihmen (1st NPP) | TPC | 6 | 3.96 |
| 3 | Hengchun (3rd NPP) | TPC | 3 | 4.5 |
| 4 | Penghu JhongtunSecond-phase | TPC | 4 | 2.4 |
| 5 | Taoyuan Datan | TPC | 3 | 4.5 |
| 6 | Taoyuan DayuanGuanyin | TPC | 20 | 30 |

---

118. http://www.moeaboe.gov.tw/About/webpage/book_en1/page5.htm.
119. Tai-Power Feed-in Tariff Ordinance (11 Nov. 2003). http://energy.ie.ntnu.edu.tw/policy/download/d48b750a63d455fccb3c8ab06e0dbe41.
120. Article 2 of Tai-Power Feed-in Tariff Ordinance (11 Nov. 2003).
121. http://web3.moeaboe.gov.tw/ECW_WEBPAGE/webpage/book_en1/page1.htm.

|   | Wind Farm | Developer | Number of Wind Turbines | Capacity (MW) |
|---|---|---|---|---|
| 7 | Taichung Power Plant | TPC | 4 | 8 |
| 8 | Taichung Harbor | TPC | 18 | 36 |
| 9 | Changbin Industrial Park | TPC | 23 | 46 |
| 10 | Hsinchu Siangshan | TPC | 6 | 12 |
| Subtotal | | | 91 | 149.76 |
| 11 | Yunlin Mailiao | Formosa Heavy Industries Corp. | 4 | 2.64 |
| 12 | Hsinchu Chunfong | Cheng Loong Corp. | 2 | 3.5 |
| 13 | Miaoli Jhunan | InfraVest GmbH | 4 | 7.8 |
| 14 | Miaoli Dapeng | InfraVest GmbH | 21 | 42 |
| 15 | Changbin Lugan | Luway | 33 | 75.9 |
| Subtotal | | | 64 | 131.84 |
| Total | | | 155 | 281.6 |

As there is limited onshore space for wind turbines and the wind is often much stronger offshore around Taiwan, there was a policy initiative to promote offshore wind power parks – The Program for the First Stage of Offshore Wind Development[122] – approved by the Executive Yuan in August 2007, with a view to developing 300 MW offshore wind farms. However, unlike the successful onshore wind farm under the feed-in tariff scheme, this program has not gone very well, and so far, no offshore wind power plants have been built. Numerous technical issues are challenging the viability of offshore wind farms, particularly concerns about the effect of typhoons and earthquakes on the wind turbine base.

As an NTD 2/kWh feed-in tariff is not very attractive to PV installers, the use of PV was encouraged by a more ambitious 'demonstration investment subsidy scheme'[123] and 'investment subsidy scheme'[124] before 2009 Renewable Energy Act. According to 2007 statistics, the Bureau of Energy subsidized the establishment of 243 PV systems sites, with a total installed capacity of 2,066 kWp, which can generate electricity of up to 2,180 kWh, enough to meet the needs of 542 households for one year. Many PV systems have been installed on public buildings for demonstration purposes, such as the Presidential Building, Legislative Yuan, and MOEA offices.

---

122. The Program of the First Stage of Offshore Wind Development, available at: http://www.moeaboe.gov.tw/opengovinfo/Laws/secondaryenergy/LSecondaryMain.aspx?PageId=l_secondary_08 (visited on 10 Oct. 2011).
123. Ordinance on Demonstration Investment Subsidy for PV Plants (enacted 6 Mar. 2002; revised on 30 Apr. 1993 and 30 May 1994).
124. Ordinance on Investment Subsidy of PV Plant (2006).

Moreover, the system is also widely installed in schools and remote areas on offshore islands for demonstration and disaster reduction purposes.[125]

There are also support measures for green procurement, that is, the development of PV units by public sector organizations. In order to deal with the financial crisis, the government sought to launch a large-scale Public Investment Program on PV Demonstration, under a special 2009 Ordinance.[126]

In addition, government is also very keen on establishing a special program to encourage the development of special forms of PV. These programs are summarized in the Table 9.18 below.

*Table 9.18   PV Programs in Taiwan since 2007*

| Promotion Projects | Project Contents |
| --- | --- |
| Solar Community 2007 ~ | PV systems for residential houses in special communities designated by local governments. 2,000 kWp will be installed by 2009. The total installation capacity will be up to 4,000 kWp. |
| Solar Top Project for each county (2007 ~ 2010) | PV systems integrated with landmark buildings, designed by local governments. |
| Solar Campus Project 2008 ~ | PV systems installed in junior high schools and elementary schools for education purposes |
| Public Building Installation Project 2009 ~ | Public construction projects with budgets over NTD 50 million must install PV systems with at least 5% expense of the total project budget being for PV installations. |

*Source*: MOEABOE, Sustainable Development of Renewable Energy, available at: http://web3.moeaboe.gov.tw/ECW_WEBPAGE/webpage/book_en1/page2.htm (visited on 10 October 2011).

Finally, there are also certain measures to encourage the development of a relatively infant technology-fuel cell. In order to encourage the initial application of the fuel cell, a Demonstration Subsidy Ordinance on Fuel Cells[127] was promulgated in January 2009. This Ordinance provides subsidies for not only the installations but also the operational fuel, as fuel cost is also a barrier to the use of fuel cells.

---

125. http://web3.moeaboe.gov.tw/ECW_WEBPAGE/webpage/book_en1/page2.htm.
126. Public Investment Program on PV Demonstration, 2009, available at: http://www.moeaboe.gov.tw/opengovinfo/Laws/secondaryenergy/LSecondaryMain.aspx?PageId=l_secondary_13 (visited on 10 Oct. 2011).
127. Demonstration Subsidy Ordinance on Fuel Cell. http://www.moeaboe.gov.tw/opengovinfo/Laws/secondaryenergy/LSecondaryMain.aspx?PageId=l_secondary_12.

## [3] Phase III: 2009 Renewable Energy Act: FIT – Adoption of Renewable Energy Act in 2009

After a long discussion in Parliament, the Renewable Energy Act was passed right after the National Energy Conference of 2009. In general, even though this Act is called 'renewable energy,' the measures mainly focus on the development of a 'feed-in tariff' scheme for 'renewable electricity.' It also reflects the fact that of all the 23 provisions, only 1 provision deals with renewable heating and renewable transportation fuel. With a view to improving the previously insufficient NTD 2 feed-in tariff, which was mainly based on the avoidance cost, the new feed-in tariff scheme will be mainly based on the real cost of each renewable electricity installation. The major measures in this Act to promote different forms of renewable energy have been illustrated in Table 9.19.

Table 9.19 Main Measures to Promote Renewable Energy under the Renewable Energy Act

| Types of Renewable Energy | Measures | |
|---|---|---|
| Renewable electricity | Main mechanisms | Renewable Energy Development Fund for R&D, investment subsidy, demonstration subsidy, and feed-in tariff |
| | | Feed-in tariff |
| | | Investment subsidy for installations and demonstration installations |
| | | Interconnection (grid connection) obligation for utilities |
| | Others | Renewable electricity target |
| | | Tax incentives for import of renewable installations |
| | | Green procurement |
| Renewable heating | Investment subsidy for solar water installations | |
| Renewable transportation fuel (biodiesel and biofuel) | Mandatory Mixture Obligation (2% for biodiesel) Voluntary program for ethanol (3% for ethanol) | |

Source: Compiled by the author.

In the second stage, the funding for the feed-in tariff (FIT) comes mainly from Tai-Power's budget. In order to provide a solid financial basis for the feed-in tariff, a Renewable Energy Development Fund was established under Article 7 of the Renewable Energy Act. The fund's resources come mainly from the non-renewable electricity generators of Tai-Power, independent power producers, and (≥300 MW) cogenerators. Government may also use its budget to finance this fund. In this regard, as there are sufficient funds for the feed-in tariff, a more preferential feed-in tariff scheme will likely be introduced. The preferential tariff for the first year, 2010 and 2011, is illustrated in the Table 9.20.

Table 9.20  Feed-In Tariff for All Renewable Electricity in 2010 and 2011

| Type of Renewable Electricity | | 2010 | 2011 |
|---|---|---|---|
| Run of River hydropower | | 2.0615 | 2.1821 |
| Wind power | | | |
| 1 kw ~ 10 kW < | | 7.2714 | 7.3562 |
| ≥10 Kw | | 2.3834 | 2.6138 |
| Offshore | | 4.1982 | 5.5626 |
| Geothermal electricity | | 5.1838 | 4.8039 |
| PV | | | |
| rooftop | 1 kW-10 kW < | 11.1883 | 10.3185 |
| | 10 kW-100 kW < | 12.9722 | 9.1799 |
| | 100 kW-500 kW < | | 8.8241 |
| | ≥500 kW | 11.1190 | 7.9701 |
| Ground-type | | | 7.3297 |
| Biomass electricity | | | |
| Bio-gas | | 2.0615 | **2.1821** |
| Waste (RDF) | | 2.0879 | 2.6875 |
| **Other (Marine energy, Fuel cell, etc)** | | 2.0615 | **2.1821** |

*Source*: Feed in Tariff Schedule of 2010, available at:http://www.moeaboe.gov.tw/Download/Policy/Renewable/meeting/files/law/law_f01%E4%B8%AD%E8%8F%AF%E6%B0%91%E5%9C%8B%E4%B9%9D%E5%8D%81%E4%B9%9D%E5%B9%B4%E5%BA%A6%E5%86%8D%E7%94%9F%E8%83%BD%E6%BA%90%E9%9B%BB%E8%83%BD%E8%BA%89%E8%B3%BC%E8%B2%BB%E7%8E%87%E5%8F%8A%E5%85%B6%E8%A8%88%E7%AE%97%E5%85%AC%E5%BC%8F.pdf; Feed in Tariff Schedule of 2011, available at: http://www.esdtaiwan.edu.tw/upload/%7BEE832687-2AD7-4D05-B91D-83CC2D3508F8%7D/%E4%B8%AD%E8%8F%AF%E6%B0%91%E5%9C%8B100%E5%B9%B4%E5%BA%A6%E5%86%8D%E7%94%9F%E8%83%BD%E6%BA%90%E9%9B%BB%E8%83%BD%E8%BA%89%E8%B3%BC%E8%B2%BB%E7%8E%87%E5%8F%8A%E5%85%B6%E8%A8%88%E7%AE%97%E5%85%AC%E5%BC%8F.pdf (visited on 10 October 2011).

Even though the original purpose of the Renewable Energy Act was to promote all types of renewable electricity, due to the failure to consider all the costs involved during the drafting of the Act, such as that of marine energy, only PV is given preferential treatment. This phenomenon led to a 'dash for PV' in 2010. The original scheduled target for 2010 was 75 MW but the accelerated development of PV resulted in 754 MW being reached, which is equivalent to the target for the years between 2015 (430 MW) and 2020 (1,250 MW).

Even though the feed-in tariff was used to replace the investment subsidy and already plays a major role in encouraging the development of renewable electricity, the demonstration investment subsidy is also being used. This subsidy supports investment in nascent technologies such as BIPV and oceanic energy.[128] As there is no relevant category on the feed-in tariff table, it is impossible to encourage these expensive renewable forms of electricity under an FIT-only system. In addition, the aforementioned Subsidy Ordinance for fuel cells is proving effective in providing demonstration subsidies for fuel cells.

The government also plays a pioneering role in PV implementation. After the aforementioned 2009 Public Investment Program for PV installations, the second wave Public Investment Program Ordinance of 2010[129] was announced after the enactment of the 2009 Renewable Energy Act. This legislation also showed government's role in promoting PV installations. Indeed, by this stage, the government was seeking to provide funding for 'local governments' to install PV units on their public buildings.

From the above, it is evident that the Taiwan government is very keen on promoting PV use after the passing of the Renewable Energy Act. The main reason is quite simple. PV is the only renewable electricity that Taiwan has the ability to export as a product around the world. In this regard, the creation of a domestic PV market would further strengthen the competitiveness of Taiwanese industry.

*[4]    Phase IV: Post-PV Boom and Post-Fukushima: New Energy Policy of November 2011 – FIT + PV Tendering*

Due to the PV boom in 2010, the Taiwanese government reformed its PV promotion scheme in 2011. The original FIT for all renewable electricity was revised to a dual-track system. In general, most types of renewable electricity and small-scale PV installations are still subject to a favorable FIT scheme. However, Taiwan incorporated a tendering system into FIT on 17 March 2011.[130]

*[a]    Eligibility*

After the PV boom, Taiwan adopted a mixed scheme of fixed FIT and bidding mechanisms. Installations are classified into four categories: Type I) rooftop installations with a capacity of 1–10 kW, when the applicant is the building owner and the roof is owned by the building owner; Type II) same as Type I, but the applicant does not own the building; Type III) ≥10 kW rooftop PV installations; and Type IV) ≥1 kW ground-mounted PV installations conforming to the land code. Generally, a minimum capacity of 1 kW is still required. In addition, there is a *cap for applications (1–2000 kW)*[131] for Types II, III, and IV.[132] Similar measures are included in the *Bidding Guideline for PV Installations Electricity Generation Installation 2011*, Ministry of

---

128. Demonstration Subsidy Ordinance on Renewable Electricity Installations. http://law.moj.gov.tw/LawClass/LawAll.aspx?PCode=J0130040.
129. Public Investment Program Ordinance of 2010. (http://solarpv.itri.org.tw/subsidy/PublicWorks/download/990330.pdf).
130. PV Tendering Ordinance of Phase I of 2011.
131. PV Tendering Ordinance of Phase I of 2011.

Economic Affairs, Phase II, but additional restrictions on systems with an installed capacity of *1 MW* are imposed on ground-mounted installations. The guideline also excludes from bidding eligibility *related government bodies whose public buildings are required to use 6% green energy, as well as installations that have ever received investment subsidies or grants from the Bureau of Energy.*

However, to reflect the need to promote renewable electricity, a new energy policy was announced in November 2011. Eligibility for applying for the FIT was broadened so that only larger rooftop PV installations (>30 kW, rather than the original 10 kW) would be subject to the tendering scheme.

[b]   Duration

Taiwan does not plan to adjust the FIT duration of 20 years for PV installations.

[c]   Rate Schedule

Due to the preferential rate of 2010, a PV boom occurred,[133] and the Taiwanese government was forced to take some countermeasures. Initially, the government attempted to greatly decrease the rate schedule for PV installations,[134] which faced a strong outcry[135] from the PV industry. However, as described above, a PV bidding scheme was introduced with four PV categories (Type I: rooftop installations with capacity of 1–10 kW, when the applicant is the building owner and owns the roof, Type II: same as Type I, but the applicants do not own the building; Type III: ≥10 kW rooftop PV installation; and Type IV: ≥1 kW ground-mounted PV installation conforming to the land code). (The PV categories, please see Table 9.21.) For Type I, the full rate applies without the need to enter the bidding process. The remaining three categories (Types II, III, and IV) must undergo the PV bidding process and are highly unlikely to receive the pre-defined rate schedule (so-called 'cap rate'). In other words, the government will develop a fixed yearly installed capacity plan and cap (12,000 kW for rooftop and 3,000 kW for ground-mounted) for installers to bid on. During the bidding process, the applied capacity for a single case cannot exceed 2,000 kW, and normally the bidder

---

132. Article 3 of PV Tendering Ordinance of Phase I of 2011: 'Eligible entities: The applicant shall apply for the PV Electricity Generation Plant according to the Art. VI of Ordinance on Renewable Electricity License and Registration for approval, and the installed capacity applied by each case shall be between 1 kW–2000 kW, and shall fulfill the following requirements: Rooftop with an installed capacity between 1 kW–10 kW and not owned by the building owner.
    Rooftop installation with an installed capacity superior to 10 kW.
    Stand-alone(ground type) installation with an installed capacity superior to 1 kW which fulfills the "Land Use Regulations".'
133. Lin, Policy U-turn of PV Promotion Scheme, 2010-12-27, *Liberty Times*, available online at http://www.libertytimes.com.tw/2010/new/dec/27/today-e8.htm.
134. Ordinance of FIT Rate Schedule of 2011.
135. *See* Yang, The 30% Rate Cut of PV Purchasing Rate: The future is dim for PV industry, 2011-03-15, *The Economy Times*, available online at http://opencongress.ccw.org.tw/law-news/1903 (last retrieved 2 Aug. 2013).

offers a bid price lower than the cap rate. Only the winners with allocated capacity (those with lower prices are prioritized) are eligible to enter into a contract.[136]

Table 9.21 Categories of PV Rate Schedule

| Renewable Energy Category | Category | Installed Capacity | Cap Rate (NT/kW) |
|---|---|---|---|
| Photovoltaics (PV) | Rooftop | 1–10 kW | 10.3185 |
| | | 10–100 kW | 9.1799 |
| | | 100–500 kW | 8.8241 |
| | | > 500 kW | 7.9701 |
| | Ground-mounted | > 1 kW | 7.3297 |

Phase I of the bidding process for PV installations has been executed, and as a result, 123 rooftop installations have been awarded the bid, with an accumulated awarded capacity of 12,173.123 kW. Discount rates range from 0% to 25%, with an average discount rate of 2.62%. For ground-mounted installations, two projects have been awarded the bid, with an accumulated awarded capacity of 1,379.4 kW. Discount rates range from 0% to 0.51%, with an average discount rate of 0.31%.[137]

Immediately afterwards, on 29 July 2011, the Taiwanese government proceeded with Phase II, Stage 1 of the PV Electricity Generation Plant Bidding Procedure; 40 rooftop installations won the bid, with an accumulated awarded capacity of 2,583.181 kW. The discount rate ranged from 1.25% to 6%, with an average discount rate of 2.95%. Only one ground-mounted installation was awarded the bid with a capacity of 248.64 kW and a discount rate of 0.31%. The total accumulated awarded capacity is 2,831.821 kW with a remaining capacity of 2,168.179 kW.[138] Therefore, only Type I can be awarded the pre-defined rate under the Decree, but the rates under the decree for the other types of PV can only be seen as a cap rate or a 'reference rate for submitting a bid.' The actual awarded rate is determined by the bidding process. A comparison of the rate schedule of 2010 and 2011 is illustrated in Table 9.22.

---

136. Ordinance of FIT Rate Schedule of 2011.
137. Bureau of Energy, Ministry of Economic Affairs: PV Electricity Generation Plant Bidding Procedure 2011 Phase I successfully completed.(29 Apr. 2011), information source: http://www.moeaboe.gov.tw/Policy/Renewable/news/SENewsDetail.aspx?serno = 01116&TYPE_KIND = News (last visited 2 Sep. 2011)ttp://www.moeaboe.gov.tw/Policy/Renewable/news/SENewsDetail.aspx?serno = 01116&TYPE_KIND = News (last visited 5 Jul. 2011).
138. MOEABOE, The bidding results of PV Tendering Ordinance of Phase II of 2011, http://www.moeaboe.gov.tw/Policy/Renewable/news/SENewsDetail.aspx?serno = 01150&TYPE_KIND = News (last visited 2 Sep. 2011).

Table 9.22 Comparison Table for the FIT Tariff for Renewable Energy in 2010 and 2011, Taiwan

| Item | | | Tariff of 2010 (A) | Tariff of 2011 (B) | Percent Change (B/A - 1) × 100% |
|---|---|---|---|---|---|
| Photovoltaic (PV) | Rooftop | 1–10 kW | 11.1883 (equivalent to 14.6030) | 10.3185 | -29.34% |
| | | 10–100 kW | 12.9722 | 9.1799 | -29.23% |
| | | 100–500 kW | | 8.8241 | -31.98% |
| | | > 500 kW | 11.1190 | 7.9701 | -28.32% |
| | Ground-mounted | | | 7.3297 | -34.08% |

Source: compiled from this article.

*[d]    Development Target and Cap*

According to Article 6(2) of the Renewable Energy Act of 2009, the *total installation capacity cap* of renewable energy development was set at 6,500–10,000 MW, but no total installation cap was specified for PV installations.

Due to the PV boom in 2010, there was substantial discussion of whether the *renewable energy promotion target* of Article 6(1) of the Act could be used as justification for introducing an annual development cap for PV installations.[139] Considering the need for administrative reliability, the Taiwanese government did not take any restrictive measures in 2010; however, it was implied that it might still be possible to establish caps under the authority of Article 6(1) in the future.

While planning the FIT promotion scheme of 2011 at the end of 2010, Taiwan was expected to limit the development of PV electricity to within 70 MW. Such measures were intended to avoid reaching the development target too quickly; the rapid installation of too many PV systems would dramatically affect the price of electricity. At that time, various measures were being considered, such as creating a bidding system or adopting a 'first applied, first reviewed' system.[140] The wording of 'first applied, first verified' implied an annual hard cap for PV.

---

139. The Executive Yuan is worried of an overheat while the target is not achieved (2010/10/10), Chinatimes.com, information source: http://news.chinatimes.com/focus/0,5243,50106873 x112010101000127,00.html (last visited: 13 Aug. 2011); purchase on PV electricity, over 35 billions is estimated, SEMI Taiwan, information source: http://www.semi.org/ch/Press/PV_news/CTR_040774 (last visited: 13 Aug. 2011).
140. price competing or first applied first verified, purchase tightened on PV electricity, Commercial Times, available at: http://money.chinatimes.com/news/news-content.aspx?id = 201012130 00018&cid = 1211 (last visited: 3 Aug. 2011).

It was strongly disputed whether Article 6 of the Renewable Energy Act could provide a legal basis for establishing an annual hard cap on development. Thus, at the beginning of 2011, the government decided to use its *Feed-in Tariff Ordinance for Managing the Application Process*[141] to officially establish a 'PV annual installed capacity hard cap.'[142] On this basis,[143] the Taiwanese government simultaneously developed a bidding system for the PV industry by publishing the *Bidding Guideline for PV Installations Electricity Generation Plant 2011, phase I, Ministry of Economic Affairs*, establishing an *annual PV hard cap* of 70 MW for 2011.[144]

Under this mechanism, the new scheme categorized PV installations into four types. Type I, the first category, included all of the self-owned, small-scale rooftop PV installations (installed on the rooftop by the residence owner with a capacity of 1–10 kW), which were favored by the government. *No annual hard cap* was created for this category. The installers are eligible for the full subsidy based on the 'renewable energy electricity FIT formula 2011'[145] and thus do not need to go through the bidding process.

For the other categories (Types II–IV), the Taiwanese government was rather discrete in its concern over a PV boom and providing an excess subsidy. The total capacity for the first bidding process, including Types II–IV, was 15,000 kW, among which the restrictions were still more *favorable for small-scale non-self-owned and large-scale rooftop installations* in comparison with ground-mounted installations. This policy is apparent in that the capacity cap allocated for rooftop installations was four times that for ground-mounted installations; the total bidding capacity for rooftop installations was 12,000 kW and 3,000 kW for ground-mounted installations.[146] Moreover, as the tariff can only be determined through the bidding procedure, the installer is not guaranteed the provisional FIT rate. As stated previously, the first stage of the bidding procedure was completed in April 2011 and the total capacity awarded was 13,552.523 kW,[147] which represented only 20% of the scheduled target for 2011 (70,000 kW); therefore, considerable growth could be expected in the future. After the second stage of the bidding procedure in 2011, in which a similar process was followed,

---

141. Ordinance on Renewable Electricity License and Registration.
142. Article 5 of the 'Ordinance on Renewable Electricity License and Registration' according to c1 of the previous article, the central administration authorities can, based upon the annual promotion target and its status of allocation, decide to accept, suspend to accept or refuse to recognize a renewable electricity installations.
143. 'In order to procede the bidding procedure on PV installations electricity generation plant, the Ministry of Economic Affairs (referred to as the Ministry) created this guideline based upon the promotion target allocation defined by the Art. 5 of the Managing Regulation of Renewable Energy Generation Installation as well as the Renewable Energy Electricity FIT Tariff and the Formula 2011, R.O.C.', Article 1 of PV Tendering Ordinance of Phase I of 2011.
144. PV installation capacity targeted at 2500 MW for 2013, (15 Apr. 2011), IDN.com, available at: http://www.idn.com.tw/news/news_content.php?catid=2&catsid=3&catdid=0&artid=20 110416abcd00 (last visited: 13 Aug. 2011).
145. Ordinance of FIT Rate Schedule of 2011.
146. PV Tendering Ordinance of Phase I of 2011.
147. PV Installations Electricity Generation Plant Bidding procedure phase I completed, Ministry of Economic Affairs, available at: http://www.idn.com.tw/news/news_content.php?catid=2& catsid=3&catdid=0&artid=20110430abcd013(last visited: 13 Aug. 2011).

the installed capacity had increased by 2,831.821 kW and 4,951.23 kW after conclusion of the first and second stages, respectively.[148]

## §9.03 ANALYSIS: UNIQUE SCHEMES IN JAPAN, SOUTH KOREA, AND TAIWAN

### [A] Comparison with German-Style FIT: Unique FITs in Japan, South Korea, and Taiwan

Before introduction of FITs, Germany's situation and approach was similar to those of Taiwan, South Korea, and Japan. For instance, before establishing a FIT, Germany sought to use other tools, such as investment subsidies, demonstration subsidies, and R&D funding, to encourage development of renewable electricity. Japan, South Korea, and Taiwan all did the same. In addition, before formal introduction of the FIT, Germany had a limited and early version of a FIT in 1991, similar to the limited FIT of Taiwan in 2003 and voluntary net-metering in Japan. In these early versions of FITs, electric utilities often played a pioneering role and were required by the government to do something to promote deployment of renewable electricity. Finally, these early efforts mainly resulted in development of wind power plants. All of these countries wanted to adopt further measures to promote development of all types of renewable electricity, but also particularly had PV in mind.

However, even though these countries shared a quite similar background before introduction of FITs and declared their intentions in their energy policies, they did not adopt ambitious versions of FIT schemes in promoting renewable electricity. Based on the results of this study, Asian-style FIT is quite different from German-style FIT in many aspects; more detailed explanation is provided below.

### [1] Eligibility

As the basic idea of a FIT is to provide the most support for development of renewable electricity, eligibility requirements are often as easy to meet as possible. For instance, under Germany's FIT, *most types* of renewable electricity are eligible for the FIT. In addition, the rate schedule is designed in a very detailed manner to reflect different sizes (and different input fuels) of renewable electricity installations.

In the FIT laws of Japan, Taiwan, and South Korea, it also appears as if most type of renewable electricity are subject to the FIT scheme. Yet, looking more closely, this may not be the case. For instance, the FITs in Japan and Taiwan have many restrictions

---

148. PV installation electricity generation plant biding, phase I stage II of 2011, 41 bids are awarded with a total bidding capacity of 2831.821 kW, http://www.moea.gov.tw/Mns/populace/news/News.aspx?kind = 1&menu_id = 40&news_id = 22239.

   PV installation electricity generation plant biding, phase II stage II of 2011, 39 bids are awarded with a total bidding capacity of 4951.23 kW, available at: http://www.moea.gov.tw/Mns/populace/news/News.aspx?kind = 1&menu_id = 40&news_id = 22239 (last visited 3 Aug. 2011).

on the renewable electricity installations that qualify for the FIT. In contrast, the FIT in South Korea is closer to that of Germany although there are fewer types of renewable electricity on the FIT list.

### [2] Duration

Germany adopted differential durations (15, 20, 30) for different types of renewable electricity installations. Japan adopted a 10–15–20 regime and South Korea a 15–20 regime, while Taiwan adopted 20 years for all types.

### [3] Rate Schedule

#### [a] Rate Schedule

A successful aspect of Germany's FIT is its multi-year rate schedule, which provides a stable investment environment for the renewable electricity industry and potential installers. Even after several-years PV booms in Germany and many European countries, the multi-year rate schedule with tariff degression was maintained.

The most similar rate schedule among the surveyed countries is that of South Korea, which remains effective until 2020. However, in Taiwan and Japan, an annual rate schedule is used, which provides less investment security.

#### [b] Tariff Degression

The FIT in Germany includes a pre-determined degression rate or target-responsive degression rate for new installers, including annual, semi-annual, and seasonal degression rates. This scheme supplements the multi-year schedule to provide a secure investment environment for renewable electricity. Among the surveyed countries, South Korea's FIT adopted a similar approach with detailed degression rates for different types of renewable electricity.

However, as noted above, this approach was not adopted by Taiwan and Japan. There is no pre-determined degression rate for late-comers. However, tariffs in subsequent years are likely to be degressed somewhat to respond to technology development.

### [4] Cost-Sharing Scheme

To support their costly FIT, Germany passes the cost on to all consumers via a system surcharge under the German Renewable Energy Act. Thus, a stable funding source for the FIT is not a concern. Among the survey countries, Japan has also adopted a comprehensive cost-sharing scheme. However, the funding schemes are relatively weak in Taiwan and South Korea.

## [5] Cap

The concept of a FIT in Germany is a program without a cap. Even if Germany introduces a cap this year, this will have resulted from multiple years PV booms. Japan has not yet adopted a cap.

In Taiwan and South Korea, there has always been a long-term cap in their Acts. Furthermore, since the PV boom, annual caps for certain types of renewable electricity have come into play.

## [6] Grid Connection and Mandatory Contracting Duty

The importance of the FIT in Germany is reflected in its comprehensive regulations on mandatory grid connections, usage rules, and mandatory contracting requirements in the German Renewable Energy Act.

However, these rules are less clear in Japan, South Korea, and Taiwan, which may give too much discretion to the electricity incumbents to the detriment of other project proponents.

## [7] Summary

Table 9.23 summarizes the above comparisons and indicates how difficult it has been for Asian countries to adopt all of the essential elements of the German-style FIT. Without a comprehensive design that includes all FIT elements, there is always a concern over sustainability. For instance, the failure of South Korea's and Taiwan's FITs may be related to the flawed design of their cost-sharing mechanisms, which led to a change in direction from the FIT to RPS and PV tendering. Due to concerns over the bankruptcy of Tai-Power in Taiwan and the financial burden on the Korean Government, the FITs in these countries have been adjusted or changed. Although the FIT in Japan is closer to Germany's design, unclear grid connection and usage rules and unclear flexibility in eligibility guidelines may present potential barriers in the future.

*Table 9.23  Comparison of FIT Elements in Germany, Japan, Taiwan, and South Korea*

|  | Germany | Japan | Taiwan | South Korea |
|---|---|---|---|---|
| Eligibility | *** | * | * | ** |
| Duration | *** | *** | * | *** |
| Rate scheme | *** | * | * | *** |
| Cost-sharing | *** | *** | * | * |
| Cap | *** | *** | * | * |
| Grid connection and usage | *** | * | * | * |

*Note*: The number of * refers to the soundness of FIT elements.

*Source*: Compiled by this author.

## [B] Comparison with German-Style FIT: Unique Schemes for PV

In Germany, incentives for PV have been provided primarily through the FIT for nearly two decades, since 1991. Although there have been certain adjustments to FIT elements (such as a target-responsive tariff degression scheme) after the PV boom, the main approach remains the FIT scheme. This reliance on the FIT reflects Germany's emphasis on both development of the domestic *PV industry* and *domestic application* of it. However, in Asian countries, a FIT for 'all' PV appears ephemeral!

In Japan, PV was subject to an RPS, then mandatory net-metering, and then a FIT combined with small PV mandatory net-metering. In Korea, PV enjoyed strong development under the FIT, but this approach was revised to a PRS with an individual PV annual capacity requirement. In Taiwan, the FIT under the 2009 Renewable Energy Act led to the PV boom and also a new mixed regime of FIT and tendering for the PV sector.

This approach of *designing a unique promotion scheme for PV* is a very interesting phenomenon among the surveyed countries. Most types of renewable electricity are subject to the primary promotion scheme in Taiwan (FIT), Japan (RPS, FIT), and South Korea (RPS), but PV has a unique status. Such uniqueness does not always indicate a preference for PV. The FIT combined with small PV net-metering in Japan and the FIT combined with tendering in Taiwan may be used as tools to discourage development of certain types of PV. The mandatory annual PV capacity introduced under the RPS scheme in South Korea also sends mixed messages: on the one hand, it appears that PV enjoys preferential status; on the other hand, the change in approach from FIT to RPS resulted from a need to curb the PV boom. Therefore, it appears that the mandatory PV net-metering scheme in Japan in 2009 was the only approach that was entirely preferential to PV.

Why such a different attitude toward PV? Perhaps it is related to the purpose of promoting PV. Compared with Germany's intent to support the development of a domestic *PV industry* and *domestic applications*, these three Asian countries face a dilemma. Promotion of domestic applications is helpful in strengthening domestic PV companies and expanding the overseas market, but heavy reliance on PV in the manufacturing sector may increase the price of electricity, production costs of PV, and the competiveness of other energy-intensive industries. Thus, the government wants to promote PV, but is also concerned about possible negative consequences. This dilemma also has led to the unique schemes for PV described here.

## §9.04 CONCLUSIONS

*Effects of Fukushima?* After the Fukushima accident, it may appear that Japan, Taiwan, and South Korea have all adopted a new renewable electricity incentive scheme. Yet, after a closer look, we find that only Japan's new FIT scheme actually responds to energy policy needs in a post-Fukushima age. The post-Fukushima measures in Taiwan and South Korea had already been drafted and proposed before Fukushima.

This relationship also affects similarity to the German-style FIT scheme. Due to the need to promote renewable electricity in Japan, Japan move from its original net-metering or RPS to a system closer to the German-style FIT to encourage large-scale domestic application of PV and other renewable electricity. Because the new schemes in Taiwan and South Korea had little or nothing to do with the Fukushima accident, both are moving away from the FIT: South Korea has abolished the FIT entirely, while Taiwan has introduced a PV tendering scheme.

*Contribution.* There exists substantial introductory literature on the development of renewable energy in the individual countries of Taiwan, South Korea, and Japan. The contribution of this article is threefold. First, Part II provides a concise and comprehensive account of the historical development of the primary promotion schemes in these three countries. Thus, readers can more easily compare the timing and nature of the development of renewable electricity promotion schemes in these three Asian countries. Second, Part III provides a comparison between European and Asian FITs. From this comparison, we have determined that Asian countries have had difficulty adopting all of the important design elements of German- or European-style FITs; the closest approach is that of South Korea. However, even this scheme did not remain in application for very long. Third and finally, via comparative evaluation, this study has brought to light that all these Asian countries have unique schemes for promoting PV that are also different from the schemes in Germany and Europe. Therefore, by comparing these three countries with a properly designed analysis structure, it is possible to find 'coherent incentive patterns' that cannot be distinguished from country-by-country introductory studies.

ENERGY AND ENVIRONMENTAL LAW & POLICY SERIES

1. Stephen J. Turner, *A Substantive Environmental Right: An Examination of the Legal Obligations of Decision-makers towards the Environment*, 2009 (ISBN 978-90-411-2815-7).
2. Helle Tegner Anker, Birgitte Egelund Olsen & Anita Rønne (eds), *Legal Systems and Wind Energy: A Comparative Perspective*, 2009 (ISBN 978-90-411-2831-7).
3. David Langlet, *Prior Informed Consent and Hazardous Trade: Regulating Trade in Hazardous Goods at the Intersection of Sovereignty, Free Trade and Environmental Protection*, 2009 (ISBN 978-90-411-2821-8).
4. Louis J. Kotzé and Alexander R. Paterson (eds), *The Role of the Judiciary in Environmental Governance: Comparative Perspectives*, 2009 (ISBN 978-90-411-2708-2).
5. Tuula Honkonen, *The Common but Differentiated Responsibility Principle in Multilateral Environmental Agreement's: Regulatory and Policy Aspects*, 2009 (ISBN 978-90-411-3153-9).
6. Barbara Pozzo (ed.), *The Implementation of the Seveso Directives in an Enlarged Europe: A Look into the Past and a challenge for the Future*, 2009 (ISBN 978-90-411-2854-6).
7. Henrik M. Inadomi, *Independent Power Projects in Developing Countries: Legal Investment Protection and Consequences for Development*, 2010 (ISBN 978-90-411-3178-2).
8. Nahid Islam, *The Law of Non-Navigational Uses of International Watercourses: Options for Regional Regime-Building in Asia*, 2010 (ISBN 978-90-411-3196-6).
9. Yasuhiro Shigeta, *International Judicial Control of Environmental Protection: Standard Setting, Compliance Control and the Development of International Environmental Law by the International Judiciary*, 2010 (ISBN 978-90-411-3151-5).
10. Katleen Janssen, *The Availability of Spatial and Environmental Data in the European Union: At the Crossroads between Public and Economic Interests*, 2010 (ISBN 978-90-411-3287-1).
11. Henrik Bjørnebye, *Investing in EU Energy Security: Exploring the Regulatory Approach to Tomorrow's Electricity Production*, 2010 (ISBN 978-90-411-3118-8).
12. Véronique Bruggeman, *Compensating catastrophe victims: A Comparative Law and Economics Approach*, 2010 (ISBN 978-90-411-3263-5).
13. Michael G. Faure, Han Lixin & Shan Hongjun, *Maritime Pollution Liability and Policy: China, Europe and the US*, 2010 (ISBN 978-90-411-2869-0).

14. Anton Ming-Zhi Gao, *Regulating Gas Liberalization: A Comparative Study on Unbundling and Open Access Regimes in the US, Europe, Japan, South Korea and Taiwan*, 2010 (ISBN 978-90-411-3347-2).
15. Mustafa Erkan, *International Energy Investment Law: Stability through Contractual Clauses*, 2011 (ISBN 978-90-411-3411-0).
16. Levente Borzsa´k, *The Impact of Environmental Concerns on the Public Enforcement Mechanism under EU law: Environmental protection in the 25th hour*, 2011 (ISBN 978-90-411-3408-0).
17. Tarcísio Hardman Reis, *Compensation for Environmental Damages under International Law: The Role of the International Judge*, 2011 (ISBN 978-90-411-3437-0).
18. Kim Talus, *Vertical Natural Gas Transportation Capacity, Upstream Commodity Contracts and EU Competition Law*, 2011 (ISBN 978-90-411-3407-3).
19. WangHui, *Civil Liability for Marine Oil Pollution Damage: A Comparative and Economic Study of the International, US and Chinese Compensation Regime*, 2011 (ISBN 978-90-411-3672-5).
20. Chowdhury Ishrak Ahmed Siddiky, *Cross-Border Pipeline Arrangements: What Would a Single Regulatory Framework Look Like?*, 2012 (ISBN 978-90-411-3844-6).
21. Rozeta Karova, *Liberalization of Electricity Markets and Public Service Obligations in the Energy Community*, 2012 (ISBN 978-90-411-3849-1).
22. Sandra Cassotta, *Environmental Damage and Liability Problems in a Multilevel Context: The Case of the Environmental Liability Directive*, 2012 (ISBN 978-90-411-3830-9).
23. Mark Wilde, *Civil Liability for Environmental Damage: Comparative Analysis of Law and Policy in Europe and US*, 2013 (ISBN 978-90-411-3233-8).
24. Bernard Taverne, *Petroleum, Industry and Governments: A Study of the Involvement of Industry and Governments in Exploring for and Producing Petroleum*, 2013 (978-90-411-4563-5).
25. Anton Ming-Zhi Gao & Chien Te Fan (eds), *Legal Issues of Renewable Energy in the Asia Region: Recent Developments in a Post-Fukushima and Post-Kyoto Protocol Era*, 2014 (978-90-411-4856-8).